Granular Activated Carbon

Design, Operation and Cost

Robert M. Clark
Benjamin W. Lykins, Jr.

 LEWIS PUBLISHERS

Library of Congress Cataloging-in-Publication Data

Clark, Robert Maurice.
    Granular activated carbon: design, operation, and cost / by
Robert M. Clark and Benjamin W. Lykins, Jr.
    p.   cm.
    Bibliography: p.
    Includes index.
    ISBN 0-87371-114-9
    1. Water—Purification—Adsorption. 2. Carbon, Activated.
I. Lykins, Ben W.   II.   Title.
TD449.5.C53   1989   89-31678
628.1'64--dc20   CIP

Second Printing 1991

LEWIS PUBLISHERS, INC.
121 South Main Street, Chelsea, Michigan 48118

PRINTED IN THE UNITED STATES OF AMERICA

Preface

This book presents a summary of design, cost, and performance information on the application of granular activated carbon (GAC) in drinking water based on field-scale experience. A brief history of the development of regulations for control of synthetic organics in drinking water is presented. The use of GAC in other countries is discussed and various design concepts for the unit operations that make up the GAC process are explored. Included in the book are chapters that present information from field-scale research projects dealing with the performance of virgin and reactivated carbon; problems and limitations of carbon reactivation systems; and use of carbon for removing trihalomethanes, trihalomethane precursors, and synthetic organics. The last chapter provides cost equations and comparative cost studies for full-scale application of GAC.

Robert M. Clark has a BS in civil engineering from Oregon State University (1969) and in mathematics from Portland State University (1961). He also has an MS in mathematics from Xavier University in Cincinnati (1964) and in civil engineering from Cornell University (1968). He received his PhD from the University of Cincinnati in 1976.

Dr. Clark, a commissioned officer in the U.S. Public Health Service since 1961, is assigned to the U.S. Environmental Protection Agency as Director of the Drinking Water Research Division, Risk Reduction Engineering Research Laboratory, Cincinnati. He is responsible for directing and conducting research that meets the technology requirements for maintaining Maximum Contaminant Levels under the Safe Drinking Water Act. This work includes formulating and guiding a broad-based program conducting research into frontier areas critically important to the provision of safe drinking water in the United States. Results of this research have had and will have a major impact on U.S. policies and procedures in water supply and on international drinking water practices.

Dr. Clark has an outstanding reputation in the government, professional, and academic communities. He is constantly sought out as a consultant and advisor by specialists and other experts, both in and out of the water supply field. Dr. Clark has received many professional awards, including selection by the American Society of Civil Engineers in 1980 as recipient of the Walter L. Huber Civil Engineering Prize. He received both the Commendation and Meritorious Service Awards from the U.S. Public Health Service (in 1977 and 1983, respectively). He has served as the chairman of the Executive Committee of the Environmental Engineering Division of the American Society of Civil Engineers and is currently vice-chairman of the Research Division Trustees of the American Water Works Association.

Benjamin W. Lykins, Jr. has a BS in chemistry from Marshall University in Huntington, West Virginia (1963) and an MS in environmental engineering from the University of Cincinnati (1974).

Mr. Lykins is currently employed at the U.S. Environmental Protection Agency as Chief of the Systems and Field Evaluation Branch, Drinking Water Research Division, Risk Reduction Engineering Research Laboratory, Cincinnati. He is responsible for providing an integrated analysis of the cost and performance of technology associated with supplying safe drinking water to the public. This work involves conducting research in the field on modular or full-scale water treatment systems, including water delivery.

Mr. Lykins has over 50 publications and presentations. He has served as a member of the American Water Works Association Academic Achievement Awards Committee and the Research Division Organic Contaminants Committee. He is also a member of several project advisory committees for the American Water Works Association Research Foundation. He is an active member of the American Society of Civil Engineers' Environmental Engineering Division Point-of-Entry/Point-of-Use Committee and the Steering Committee for the Ontario Ministry of the Environment trace organics contaminant removal project.

Contents

Introduction

The first water-related regulation — the prohibition of the use of the common drinking water cup on interstate carriers — was adopted in 1912. Subsequently, additional regulations were adopted and updated until 1974, when the Safe Drinking Water Act (SDWA) was passed. The Safe Drinking Water Act has radically changed the way in which we view the delivery of water in the United States.

Originally enacted in 1974, the Safe Drinking Water Act requires the U.S. Environmental Protection Agency (EPA) to determine which contaminants threaten public health and to set standards for those contaminants. States, aided by federal grants, enforce drinking water rules and supervise the nation's water suppliers. Suppliers must monitor their supplies for contaminants and treat the water to meet the standards.

The drinking water standards set by EPA under the act apply to public water systems (systems that have at least 15 service connections or regularly serve at least 25 individuals). It is estimated that there are more than 200,000 public water systems in the nation, including nearly 60,000 community systems.

The act also requires EPA to institute rules governing the disposal of wastes by underground injection and provides authority for grants to states to operate enforcement programs. The rules cannot impede oil and gas operations unless essential to protect underground drinking water. In areas that overlie a sole source aquifer — an underground water supply used as the principal drinking water source for an area — EPA can prohibit new injection wells or withhold federal money from projects that threaten the aquifer.

In 1986, Congress amended the SDWA with the objective of speeding EPA's efforts to set standards for chemical contaminants that have been detected in drinking water and to establish a national (though limited) groundwater program by requiring states to protect groundwater tapped by public water system wells.

EPA has identified hundreds of synthetic organic chemicals, heavy metals, pesticides, and other pollutants in drinking water supplies, usually at trace levels. Although the health effects of many chemicals are not well known, some at sufficient levels can cause eye and skin irritation, birth defects, cancer, and damage to the nervous system, liver, kidneys, and lungs. The contaminants come from pollution, from disinfection of drinking water supplies with chlorine (which can react to form dangerous by-products), and from nature.

In 1983, EPA's Ground Water Supply Survey found that about 28% of randomly selected large municipal water systems (serving more than 10,000 people) had detectable levels of one or more volatile organic chemicals in their groundwater supplies. EPA's Community Water Supply Survey found that water from 45% of large public water systems relying on groundwater had detectable levels of volatile organic chemicals.

In fiscal year 1985, EPA reported about 16,000 violations of drinking water standards and 79,000 violations of monitoring and reporting requirements. Of the 58,530 community water systems reporting, 728 were persistent violators of the microbiological standards. Most other violations were caused by exceeding the nitrate and fluoride standards.

Numerous studies have documented that most drinking water quality problems in the United States are associated with small water utilities (less than one million gallons of water supplied per day). It is likely that many of these small utilities will be required to make extensive investments in technology to meet requirements of the SDWA.

The Safe Drinking Water Act and the Safe Drinking Water Act Amendments (SDWAA) of 1986 require that public water systems use the best available technology to remove contamination and to monitor for chemicals that are not regulated. The amendments strengthen enforcement of drinking water regulations and rules for disposal of waste through underground injection wells. Specific provisions of the 1986 amendments are as follows:

• Set deadlines for EPA to regulate dangerous chemicals and other contaminants in drinking water provided by public water systems. EPA must issue standards within three years for 83 contaminants previously identified by EPA as candidates for regulation. EPA can substitute up to seven contaminants for listed contaminants if regulating the substitutes would better protect public health. The contaminants include volatile organic chemicals such as benzene, organic chemicals such as cyanide, and radioactive substances, bacteria, and viruses. For other contaminants that may require regulation, EPA established a priority list on January 1, 1988. Lists are to be established subsequently at three-year intervals. EPA must regulate those contaminants that may have any adverse effect on health.

- Require drinking water standards to be set based on best available technology, taking costs into account. Standards for synthetic organic chemicals must be based on use of technologies at least as effective as granular activated carbon.
- Require EPA to issue regulations requiring filtration or equally protective measures by public water systems that rely on rivers or other surface water sources. EPA also must issue regulations requiring disinfection for all public water systems and specifying criteria for variances. Filtration and disinfection remove bacteria and viruses.
- Require EPA to issue regulations requiring public water systems to monitor for unregulated contaminants.
- Ban the use of pipes, solder, and flux containing more than small amounts of lead in the installation or repair of public water systems. The ban also applies to plumbing in residences and other buildings connected to the systems. States must enforce the ban through state or local plumbing codes or other means.
- Require EPA to take enforcement action in every case in which drinking water or underground injection control requirements are violated if the state does not act. EPA is given authority to issue administrative compliance orders as well as take violators to court.
- Establish a nationwide program requiring states to establish "wellhead protection areas" around public drinking water system wells where pollutants could flow down into underground formations containing water supplies. Within three years, states must develop plans to prevent contamination from potential pollution sources. States are to decide the extent of each wellhead area and how to protect it. A state could determine that technical and financial assistance, education and training, or demonstration projects, without a regulatory program, would protect an area.
- Establish a demonstration assistance program for state or local entities to protect aquifers designated by EPA as the sole or principal drinking water source for an area.

The administrator is required to set Maximum Contaminant Level Goals (MCLGs) that are health-based. Each MCLG must be set by EPA at a level at which no known or anticipated health effects occur, allowing an adequate margin of safety. Each Maximum Contaminant Level (MCL) promulgated under SDWAA must be promulgated simultaneously with the publishing of the MCLG and must be set as close as is feasible to the MCLG.

In this regard, "feasible" means with the use of the best technology (treatment technique) and other means available, taking cost into consideration. In setting the MCLs for synthetic organic chemicals (SOCs), the use of granular activated carbon (GAC) for SOC control is considered feasible according to the 1986 amendments. Any other technology, treatment technique, or other means found to be the best available for the control of SOCs must be as effective as GAC for this purpose.

In addition to the determination that the use of GAC for controlling SOCs is considered a feasible treatment technique, the amendments require that for each National Primary Drinking Water Regulation (NPDWR) that establishes an MCL, the Administrator of the EPA must list the technology, treatment technique, and other means that he or she determines are feasible for meeting the MCL. There is no requirement that these means be used for meeting the MCL.

The result of these amendments is to establish GAC as a baseline technology for removal of synthetic organic chemicals. Therefore, this book has been written to provide the best and most current information available in the United States on the cost and performance of granular activated carbon.

CHAPTER 1

Drinking Water Contamination and GAC Treatment

INTRODUCTION

Concern over the quality of drinking water is not new. One of the first historical references to drinking water standards is 4000 years old, recorded in Sanskrit:

> It is directed to treat foul water by boiling it, exposing it to sunlight, and by dipping seven times into it a piece of hot copper, then to filter and cool in an earthen vessel. The direction is given by the God who is the incarnation of medical science.[1]

Waterborne disease was prevalent in the United States by the end of the nineteenth century; for example, in the 1880s, the typhoid death rate was 158 deaths per 100,000 people in Pittsburgh in one year. However, the national typhoid death rate was 5 per 100,000 in 1935. Reductions in waterborne disease outbreaks were brought about by the use of sand filtration, chlorination-disinfection, and the application of drinking water standards. Recent epidemiological evidence shows that the number of deaths from all waterborne disease outbreaks dropped from 45 per 100,000 in 1938–40 to 15 per 100,000 in 1966–70. The average annual number of outbreaks ceased to fall around 1951 and may have increased slightly since then for reasons unknown at this time.[1]

The federal authority for drinking water standards originated with the Interstate Quarantine Act of 1893, which authorized the surgeon general to make and enforce regulations to prevent the introduction, transmission, or spread of communicable diseases in the United States.[1] A provision in the

1

Table 1. Drinking Water Standards

Date of Standard	Major Feature
1914	The first bacteriological performance standard was introduced.
1925	Standards were modified to include chemical standards.
1942	Bacteriological samples were taken from distribution systems.
1946	Standards were made generally applicable to all water supplies in the United States.
1962	Requirement was made that water supply system be operated by qualified personnel. Recommended maximum concentration limits were added for: alkyl benzene sulfonate barium cadmium carbon chloroform extract cyanide nitrate silver Temperature was considered in setting limits for fluorides; section on radioactivity was added. Rationale used by the standards committee in setting limits was added.

act resulted in the promulgation of the Interstate Quarantine Regulations of 1893. In 1912, the first water-related regulation—the prohibition of the use of the common drinking water cup on interstate carriers—was adopted. Since that time a series of drinking water regulations has been issued. The dates and major implications of the regulations are shown in Table 1.[1]

A set of drinking water standards was issued in 1962. In 1969, a technical committee was established to write a new set of standards. The committee's work was completed in 1972, but the new standards were held up pending the passage of the Safe Drinking Water Act (SDWA) of 1974.

ORGANIC CONTAMINATION

Much of the recent concern over drinking water quality has been directed toward organic contamination in drinking water. This concern involves not only source water contamination but the formation of disinfection by-products resulting from the interaction of disinfectants and precursors in the source water.

In 1956, while developing the carbon filter technique of sampling for organics in drinking water, Middleton and Rosen reported on the level of organic materials in the Ohio River, a source of drinking water for many communities.[2] In 1962, a recommended limit was set at 200 mg/L of carbon chloroform extract (CCE) to prevent the "unwarranted dosage of the water consumer with ill-defined chemicals."[3]

In 1974 two papers were published that heightened interest in the subject of drinking water in the United States and The Netherlands. These papers reported that the use of chlorine as a disinfectant increases the concentration of certain halogenated organics in drinking water. Of the organic compounds created during chlorination, trichloromethanes, commonly called chloroform, are formed in the greatest quantities.[4,5]

A national survey of 80 water supplies found the universal problem of chloroform contamination following chlorination. As part of that study an attempt was made to characterize as completely as possible the purgeable organic chemicals in five typical water supplies. A total of 72 compounds was identified in the drinking water of the five cities.[6,7] Concurrent with this work, the National Cancer Institute (NCI) completed a study on chloroform as a carcinogen, indicating that chloroform was a carcinogen in test samples.[8]

In 1976 Page, Harris, and Epstein purported to establish the relationship between drinking water and cancer mortality in Louisiana.[9] Eventually, 86 specific organic chemicals were identified in New Orleans drinking water.[10]

As a reflection of this concern over the effects of organic contamination, the U.S. Environmental Protection Agency (EPA) issued an Advanced Notice of Proposed Rulemaking (ANPRM) for organic chemical contaminants in drinking water. This rule proposed regulations to deal with the control of chloroform and synthetic organics in drinking water.[11]

Sources of Contamination

Organic contamination has been thought to primarily affect surface waters, which have become the repository for the waste discharges of numerous industrial facilities and municipalities as well as urban and agricultural runoff. These surface supplies are also frequently sources of drinking water.

At the time of the passage of the Safe Drinking Water Act of 1974, more than 12,000 chemical compounds were known to be in commercial use, with many more being added each year. The causes of synthetic organic chemical contamination are chronic and variable in nature. Table 2 lists some major contributors and their possible sources and methods of introduction into drinking water.[12]

Industrial discharges from point sources are regulated by the Federal Water Pollution Control Act Amendment's National Pollutant Discharge Elimination System (NPDES). Despite this "control" of industrial discharges through the nationwide permit system, some toxic pollutants continue to be discharged into surface waters. There is always the possibility of accidental or deliberate spills. Additionally, there are any number of nonpoint sources that may contribute highly toxic pesticides; stormwater runoff carries other potentially harmful substances.[13]

Table 3 lists some of the organic chemicals considered carcinogens and mutagens found in drinking water from an 11-city survey.[14]

Table 2.   Sources of Carcinogenic Chemicals Found in Water[12]

Class of Compound	Source	Method of Introduction
Petrochemicals		
Oil(s)	Refinery waste,	Waste discharge, spills
Polycyclic aromatics	petrochemical plants,	
Methylated naphthalenes	service stations, metal	Rain, runoff, direct
Kerosene	working plants, ships,	application, spills
etc.	carriers for pesticides,	
	asphalted roads	
Coal tar		
Coal tar	Coke ovens, tar distilleries,	Waste discharge, spills
Pitch	tar paper plants, wood	
Creosote	treating, gas plants	
Anthracene		
Aromatic hydrocarbons		
etc.		
Aromatic-amino and		
-nitro compounds		
Benzidine	Dye and rubber plants,	Waste discharge, spills
4-Aminodiphenyl	pharmaceutical plants,	
Beta-naphthylamine	textile dyeing operations,	
etc.	plastics plants	
Pesticides		
DDT	Manufacturing operations,	Waste discharge, spills,
Dieldrin	use of pesticides	rainfall, runoff, settling
Aramite		from air
Carbon tetrachloride		
Acetamide		
etc.		

Table 3.   Possible Carcinogens and Mutagens Found in 11-City Survey[14]

Compounds	
Benzene	Tetrachloroethylene
Carbon tetrachloride	Trichloroethylene
Bis(2-chloroethyl)ether	Vinyl chloride
Chloroform	Bromodichloromethane
1,2-Dichloroethane	Chlorobenzene
Dieldrin	Chloromethylether
DDT, DDE	Dibromochloromethane
Heptachlor	1,3-Dichlorobenzene
Hexachlorobenzene	Dichloromethane
Lindane (γ-BHC)	Methylene chloride
PCB	Vinylidene chloride

Groundwater Contamination

In addition to surface water contamination, there is a growing awareness and concern over groundwater contamination. More than 100 million Americans depend on groundwater as a source of drinking water. Three-quarters of U.S. cities get their water supplies totally or in part from groundwater, and one-third of the largest cities rely on groundwater for at least part of their potable water supply.

A number of recent studies have documented the actual or potential contamination of our nation's groundwater on a wide scale. A survey conducted by the EPA has documented that 22% of approximately 466 randomly sampled utilities, using groundwater as their source, produced drinking water containing volatile organic chemicals (VOCs) at detectable levels.[15]

In a given well, one or two compounds will predominate at relatively high concentrations, with several other compounds being present at much lower concentrations.[16] An affected supply typically contains several VOCs; trichloroethylene (TCE) was detected most frequently and in highest concentration. A 1980 Library of Congress report identified 1360 well closings based on 128 incidents of toxic groundwater contamination from 1950 to 1980. Virtually every state has reported abandonment of drinking water supplies due to chemical contamination. TCE has led to closures of water supply wells on Long Island, as well as in other parts of New York, in Massachusetts, and throughout California. Though concern in the past five years has centered on VOC contamination of groundwater, several studies have shown the presence of pesticides in groundwater, and in recent years concern has been growing over contamination of groundwater by other agricultural chemicals (fertilizers, insecticides, nematocides, and fungicides). The concern is greater, of course, when specific chemicals are identified in homeowners' drinking water wells.

Over the last eight years, New York's Suffolk County Health Department has examined the groundwater underlying Long Island for agricultural organic constituents and their decay products.[17] During the testing period, 101 agricultural/organic compounds were evaluated, with 41 found in the groundwater. Many of these identified contaminants were present in trace quantities, but aldicarb, carbofuran, 1,2-dichloropropane, and 1,2,3-trichloropropane were at elevated levels. In addition, nitrates from fertilizer applications were present in quantities exceeding the primary drinking water standards of the Safe Drinking Water Act. In another study, conducted by the California Department of Food and Agriculture in 1983, dibromochloropropane (DBCP), ethylene dibromide (EDB), and simazine were found in San Joaquin Valley wells.[17] A report submitted to the California State Water Resources Board states that more than 2000 wells statewide have been found to be contaminated with DBCP.[18]

It is against this history of concern that the Safe Drinking Water Act was passed and amended in 1986. The following section describes the SDWA and its amendments.

THE SAFE DRINKING WATER ACT

In 1974 Title 14 of the Public Health Service Act, entitled "Safety in Public Water Systems," commonly called the Safe Drinking Water Act (PL 93-523) was passed. It applies to all public water systems as defined in the act, including federally owned or maintained water systems.[19]

The act defines contaminants, maximum contaminant levels (MCLs), primary drinking water regulations, secondary drinking water regulations, public water systems, supplies, agency heads, and so forth. Some of the definitions are as follows:

- contaminant—any physical, chemical, biological, or radiological substance or matter in water
- maximum contaminant level—the maximum concentration of a contaminant allowable in water delivered to a user
- public water system—a water system that has at least 15 connections and serves at least 25 individuals

The more than 200,000 public water systems specifically included under PL 93-523 serve cities, towns, communities, Indian communities, and federal facilities. Public water systems include those in service stations, hotels, and other facilities where the public has access to water (noncommunity systems). Of primary interest to the EPA are the approximately 60,000 community systems.

Water systems not covered under the SDWA include those that consist only of distribution and storage facilities, those that obtain all water from another public water system, those that do not sell water to any person, and those that do not supply an interstate carrier that conveys passengers.

Primary drinking water regulations specify contaminants that may have an adverse effect on health. They define a maximum contaminant level or treatment technique and contain criteria and procedures to assure a safe supply. Secondary drinking water guidelines specify maximum contaminant levels that are necessary to protect the public welfare and include such items as taste, odor, and appearance. They may vary according to regional circumstances. They are not federally enforceable, but are guidelines. EPA is authorized to make program grants if the state is willing to establish a program according to the act's provisions.

A much-neglected area included in the act is undergound water source protection, a provision that is sorely needed. EPA can award program grants to those states that will establish programs for receiving a grant and that will assume primary enforcement responsibility for the underground water regulations. The state must establish a permit program in which each applicant will assure the state that underground injection will not endanger drinking water. The program requires inspection, monitoring, and record-keeping in accordance with EPA requirements. The underground water regulations apply to federal agencies and property owned or leased by the federal government. These regulations cannot interfere with underground injection of brine associated with oil and natural gas production, or with the secondary or tertiary operations of oil and gas production.

EPA is authorized to conduct research on methods to identify and measure contaminants, treatment techniques, and protection of underground water sources, and is required to conduct studies on the costs of carrying out primary regulations and on the costs of groundwater protection techniques. EPA is authorized to make grants for the purpose of training water-treatment plant operators, inspectors, and supervisory personnel.

The act requires assurance of chemical supplies. Personnel in a public water supply facility may petition the EPA administrator to issue certification of need for water treatment chemicals. The President or his delegate can then issue an order to supply the chemicals.

The act has a public notice requirement under which a utility must give notice to its consumers if it fails to meet a primary drinking water regulation, fails to perform required monitoring, has a variance or exemption (the possibility of variances or exemptions is provided for in the act), or fails to comply with a schedule for a variance or exemption. Notice must be given every three months in a newspaper of general circulation or on water bills. In the act, there is also a citizens' civil action provision under which any citizen may commence a civil action against any person for alleged violations of the act. No suit may be brought until 27 months after enactment.

Another important part of the act is the emergency powers and provisions under which the EPA administrator may take action in case a public water supply is contaminated, if the contamination presents an imminent and substantial threat to public health. The act contains provisions for the administrator to take various courses of action when a contaminant is detected for which no MCL has been established. The control of organic contamination in drinking water is a major activity under the Safe Drinking Water Act.

Control of Organics

Four statements in the *Federal Register* trace the regulatory history of the control of organic contaminants in drinking water. The first statement was an Advance Notice for Proposed Rulemaking, published July 14, 1976, in which the EPA proposed several options for the control of trihalomethanes (THMs) and other organic contaminants in drinking water and requested comments from the public.[20]

As a result of this statement and the public comments, EPA proposed a two-part regulation for organic contaminant control in drinking water on February 9, 1978. This document specified an MCL for the control of trihalomethanes (THMs) and specified that a treatment technique, granular activated carbon (GAC), be installed at water utilities for the control of other synthetic organic contaminants where source waters were significantly contaminated.

On July 6, 1978, the EPA published supplemental technical information on this subject and extended the public comment period. The final regulation, for the control of trihalomethanes only, was published in the *Federal Register* on November 29, 1979. In this same document, the proposed regulation of other synthetic organic contaminants by using granular activated carbon treatment was deferred, to be reproposed at a later time. The trihalomethane regulation is summarized in Table 4.

On January 11, 1980, the American Water Works Association, together with the City of Englewood, Colorado, and the Capital City Water Company, a Missouri corporation, filed a petition for review with the U.S. Court of Appeals for the District of Columbia Circuit, asking the court for "a review of a final rule" as allowed by Section 1448(a)(1) of the Safe Drinking Water Act.

The proposed treatment regulation is discussed in the following section.

Proposed Treatment Regulation

The second part of the regulation was concerned with the control of source water synthetic organic chemicals (SOCs) and proposed the use of GAC adsorption as a treatment technique to be used by those water utilities using source waters that were vulnerable to possible contamination by SOCs.[21]

The EPA was proposing GAC adsorption as the mandated treatment technique only for those waters that were deemed vulnerable to synthetic organic contamination, estimated to be fewer than 50 utilities.

Three criteria were proposed for the design and operation of the GAC adsorption systems: (1) the design of the GAC system had to be such that the initial removal of total organic carbon (TOC) was 50% or greater; (2) the GAC should be replaced when the TOC concentration in the effluent had risen more than 0.5 mg/L over the initial concentration in the effluent; and (3) the GAC should be replaced when the concentration of any specific chlorinated compound (except THMs) appeared in the effluent at a concentration greater than 0.5 μg/L.

The first criterion was proposed to avoid the use of very shallow GAC beds and to make the second criterion operational. The second criterion was proposed because an increase of 0.5 mg TOC/L would indicate that organic compounds were breaking through the adsorber bed (thus, the concurrent

Table 4.   Summary of Trihalomethane Regulation (Promulgated November 1979)[19]

Maximum contaminant level
   0.10 mg/L total trihalomethanes (TTHM)

Applicability
   Community water systems that add disinfection to the treatment process (ground and surface)

Effective:

Systems > 75,000	2 years after promulgation
Systems 10–75,000	4 years after promulgation
Systems < 10,000	Discretion of primacy agency (state, or EPA if the state does not have primacy)

Monitoring requirements
   Running 12-month average of a minimum of 4 samples per quarter per treatment plant taken on the same day. Systems using multiple wells drawing raw water from a single aquifer may, with primacy agency approval, be considered to have one treatment plant for determining the required number of samples.

Effective:

Systems > 75,000	1 year after promulgation
Systems 10–75,000	3 years after promulgation
Systems < 10,000	primacy agency discretion

Sample locations:
   25% at extremity of the distribution system; 75% at locations representative of the population distribution

Adjustment of sampling frequency
   For groundwater systems, reduced monitoring may be appropriate for certain systems. The primacy agency may reduce the requirements through consideration of appropriate data, including demonstration by the system that the maximum total trihalomethane potential (MTP) is less than 0.10 mg/L. The minimum frequency would be one sample per year for MTP taken at the extremity of the distribution system. For groundwater systems not meeting the above MTP limit and for surface water systems, the primacy agency may reduce the monitoring requirements if, after one year of data collection, TTHM concentrations are consistently below 0.10 mg/L. The minimum frequency would be one sample per quarter per treatment plant for total trihalomethanes taken at the extremity of the distribution system. The original frequency would be reinstated if the TTHM concentrations exceed 0.10 mg/L or if the treatment or source is modified.

Reporting requirements

To primacy state:
   Average of each quarterly analysis, within 30 days. Until primacy states have adopted the regulations, reporting will be to EPA unless primacy state requests receipt of data.

To public and primacy state
   Running 12-month average of each quarterly sample if total trihalomethane concentration exceeds MCL, as prescribed by the public notification provisions

Other requirements

To ensure microbiologic quality:
   Primacy agency approval of significant modifications in the treatment process for the purpose of meeting the TTHM MCL

Analytic requirements:
   In accordance with specified methods (purge and trap or liquid-liquid extraction) conducted by certified laboratories

breakthrough of potentially dangerous organics was possible). Because the breakthrough of TOC occurs before that of almost any other specific organic compound, TOC breakthrough is a good indicator of the deterioration of adsorber performance. The third criterion was proposed to control chlorinated organics to the limit of detection of the analytic procedure. Many of these substances were known or suspected human carcinogens. EPA recognized that these criteria were strict but believed that their use would ensure improved water quality for the consumer.

There was a significant negative reaction to the treatment proposal. Even the National Drinking Water Advisory Council, a citizens' advisory group established by the SDWA, voted not to support the concept of a mandated GAC adsorption technique.[22]

The council recommended an alternative approach to EPA for the regulation of other synthetic organic chemicals. The council's alternative approach was intended to provide the immediate controls, through MCLs, that appeared warranted by the legislative mandate under the Safe Drinking Water Act. It allowed the application of the same justification that enabled the council to support the establishment of an MCL for THM, i.e., the results of animal studies. It provided for further animal study research, which the council believed was required to justify establishing a mandated treatment technique for control of gross synthetic organic contaminants in drinking water supplies. In addition, it allowed time to complete prototype studies of treatment techniques, which seem to be universally recognized as necessary, before a final treatment requirement was mandated. In the opinion of most members of the council, this alternative approach seemed logical and rational and would not significantly delay EPA's timetable for initiating control of organic contaminants in drinking water.

As a result of the generally negative reaction to a GAC treatment requirement, the EPA developed a plan to study the use of GAC to control organic contaminants in drinking water. According to the plan, the performance of GAC as a sand replacement medium was to be compared with the performance of adsorption following sand filtration in Cincinnati, Ohio and Jefferson Parish, Louisiana. The effectiveness of fluid bed and infrared reactivation was also to be studied. The use of powdered activated carbon as a treatment technique was to be studied in New Orleans, Louisiana, and researchers in Evansville, Indiana were to investigate the use of chlorine dioxide in conjunction with GAC adsorption. Results from these projects, in association with EPA-sponsored projects at Philadelphia, Pennsylvania and Miami, Florida were to help form the basis for possible future regulatory action. A more recent project at Wausau, Wisconsin was initiated to study the potential for using adsorption models for full-scale design of GAC adsorbers. Another project initiated at Wausau compared the use of aeration to GAC for removing volatile organic chemicals from groundwater.

Despite these early studies and a high degree of interest in their results, only a few revisions of existing regulations were made after the initial passage of the SDWA, although the SDWA requires a review of the regulations at least every three years. Congress began to place pressure on EPA to increase its regulatory activity.

In the October 5, 1983 issue of the *Federal Register*, the EPA outlined its approach to revising the National Primary Drinking Water Regulations (NPDWR). The EPA stated its intention to develop these revised regulations in five phases as follows:

    I. volatile synthetic organic chemicals (VOCs) regulations
   II. other synthetic organic chemicals (SOCs), inorganic chemicals (IOCs), and microbiological contaminants
 III. radionuclides
 IV. disinfectant by-products including trihalomethanes
  V. SOCs, IOCs, and microbiological contaminants not regulated during Phase II

New amendments to the SDWA were signed into law on June 16, 1986.[23]

1986 SDWA Amendments

The major aspects of the 1986 amendments to the Safe Drinking Water Act include:

- compulsory revisions to the drinking water regulations in a timely fashion for new contaminants
- definition of a treatment technique for each contaminant regulated
- requirement of a treatment technique where it is infeasible to ascertain the level for those regulated contaminants in water
- filtration requirement for surface water supplies with certain exceptions
- disinfection of all water supplies
- prohibition of use of lead products in all conveyances for drinking water
- requirement for protection of groundwater sources by states through wellhead protection regulations
- maintenance of a chlorine residual throughout the distribution system

Maximum Contaminant Level Goals

The 1986 amendments to the SDWA have redefined Recommended Maximum Contaminant Levels so that they are now known as Maximum Contaminant Level Goals (MCLGs). In the future MCLs and MCLGs must be proposed simultaneously and promulgated simultaneously.

Table 5. Primary Maximum Contaminant Levels for VOCs

Contaminant	Level (mg/L)
Trichloroethylene	0.005
Carbon tetrachloride	0.005
1,2-Dichloroethane	0.005
Vinyl chloride	0.002
Benzene	0.005
para-Dichlorobenzene	0.750
1,1-Dichloroethylene	0.007
1,1,1-Trichloroethane	0.200

Contaminants to be Regulated

The amendments recognize 83 contaminants for which regulations must be developed. Of these, 14 VOCs were addressed in the ANPRM of March 4, 1982. The remainder were addressed in the ANPRM of October 5, 1983. Of the remainder, 29 are new SOCs, 13 are new IOCs, 4 are new microbiological contaminants, and 2 are new radiological contaminants. Those 21 contaminants contained in the Interim Primary Drinking Water Regulations (IPDWR) are included in the total number of contaminants addressed by the amendments and were also addressed in the ANPRMs noted above. The 1986 amendments have upgraded the previous IPDWR to National Primary Drinking Water Regulations.

With regard to the 83 listed contaminants, the amendments required the EPA administrator to publish MCLGs and promulgate NPDWR (including MCLs as appropriate) for not less than nine of the listed contaminants (as contained in the two ANPRMs) within one year of enactment of the amendments. This promulgation took place and included the following VOCs: benzene, vinyl chloride, carbon tetrachloride, 1,2-dichloroethane, trichloroethylene, 1,1-dichloroethylene, 1,1,1-trichloroethane, p-dichlorobenzene, and tetrachloroethylene. Table 5 includes the MCLs for these contaminants. Another 40 of the listed contaminants must be similarly regulated within two years of enactment. Of these 40 contaminants, undoubtedly some of the 21 contaminants previously listed in the IPDWR will be included with revisions.

The remainder of the listed contaminants must be regulated as those above within three years of enactment of the amendments. Up to seven different contaminants other than those listed may be substituted if the EPA administrator finds these may take precedence as public health concerns.

Each MCLG must be set by the EPA at a level at which no known or anticipated health effects occur allowing an adequate margin of safety. Each MCL promulgated simultaneously with the publishing of the MCLG must be set as close as feasible to the MCLG.

Treatment Techniques

In this regard, "feasible" means with the use of the best technology (treatment techniques) and other means available, taking cost into consideration. In setting the MCLs for synthetic organic chemicals, the use of granular activated carbon for SOC control is considered feasible according to the 1986 amendments. Any other technology, treatment technique, or other means found to be the best available for the control of SOCs must be as effective as GAC for this purpose.

In addition to the determination that the use of GAC for SOC control is considered a feasible treatment technique, the amendments require that for each NPDWR that establishes an MCL, the EPA administrator must list the technology, treatment technique, and other means that he or she determines are feasible for meeting the MCL. These means do not have to be used for meeting the MCL.

In the event that it is not economically or technologically feasible to ascertain the level of a regulated contaminant, the EPA administrator is authorized to require the use of treatment in lieu of an MCL. The administrator must identify the treatment techniques that would prevent known or anticipated health effects. A variance may be granted from the use of the identified treatment techniques if it can be shown that an alternative technique is at least as effective. In the event a variance is granted, the treatment technique must be implemented.

The result of these amendments is to establish GAC as a baseline technology for removal of synthetic organic chemicals. In many ways, the amendments accomplish by legislative action what EPA failed to accomplish with its original combined trihalomethane and treatment rule. A brief background on the use of GAC in the United States is described in the following section.

GRANULAR ACTIVATED CARBON

In this section, U.S. and European experience is explored. Since the February 1978 EPA proposal, much has been written both for and against the use of GAC for treating drinking water. Available information on the use of GAC treatment in water and wastewater applications is summarized in this section.[24-28]

Powdered activated carbon (PAC) has been used for more than 50 years to remove taste and odor from public water supplies, but GAC has been used by only a few facilities in the United States for removing taste and odor from drinking water.

U.S. Experience

Although the use of GAC in municipal water treatment has been limited, GAC has been used in industrial and municipal wastewater treatment and in various industrial process applications. The specific uses of GAC for these applications are somewhat different from those for water treatment, but much of the information on design and operations should prove useful to water purveyors. In general, the application of GAC adsorption to drinking water treatment is simpler than its application to wastewater treatment. In order to document the use of GAC in the United States, a special study was conducted in which many water, wastewater, and industrial GAC systems were visited. Eighteen of the 22 GAC installations visited during the study were industrial process or municipal wastewater facilities, whereas four installations were municipal water treatment plants.[28] Design, operating, performance, and cost information for these installations are summarized in Table 6.

A principal function of this study was to collect operating and maintenance (O&M) and construction cost information on GAC installations.[28] Plant records were used to obtain available construction costs, the dates for these costs, and the most recent O&M costs. Although the basic data are presented in Table 6, no attempt has been made to update construction or O&M costs to present-day prices or to extrapolate the costs to water treatment plants. However, procedures are available with which data from existing GAC projects can be adjusted or modified (based on the results of pilot-plant tests of the water to be treated). These data can be used by experienced professionals in making preliminary estimates of costs for future potable water projects involving GAC adsorption and reactivation or replacement.

Although many successful applications of GAC in advanced waste treatment plants exist, the GAC experience in some advanced waste treatment plants has unfortunately been poor. Failures in advanced waste treatment applications have not stemmed from deficiencies in the basic GAC processes or in organics adsorption and thermal reactivation, but rather from mechanical problems.[29]

Operational Problems

In discussing the operational problems encountered with wastewater GAC systems, it is necessary to distinguish those problems associated specifically with sewage from general problems that might be encountered with any type of GAC system. Many problems with GAC in wastewater treatment will not occur in water purification. For example, in water treatment, few or no problems would be expected with excessive slime growths, hydrogen sulfide gas production, or corrosion from adsorbed organics released during carbon reactivation.[30]

Table 6. Summary of GAC System Characteristics

Case Number	Owner	Type of Facility	Flow mL/d	Flow mg/d	Pretreatment	Contract Time (min)	Carbon Contractors Hydraulic Loading mm/s	Carbon Contractors Hydraulic Loading gpm/sq ft	Furnace Rate Capacity kg carbon/d	Furnace Rate Capacity lb carbon/d
1	South Tahoe	Municipal wastewater	28	7.5[a]	Extensive	17	4.4	6.5	2,720	6,000
2	Tahoe-Truckee	Municipal wastewater	18	4.83[a]	Extensive	20			1,740	3,840
3	Upper Occoquan	Municipal wastewater	57	15.0[e]	Extensive	22	5.7	8.4	5,440	12,000
4	American Cyanamid	Chemprocess	76	20.0[e]	Extensive	30	5.4	8.0	55,300	122,000
5	Vallejo	Municipal wastewater	49	13.0[e]	Moderate	25	4.1	6.0	13,200	29,000
6	Orange County	Secondary effluent	57	15.0[b]	Extensive	34	3.9	5.8	5,440	12,000
7	Niagara Falls	Municipal with significant industrial	180	48.0[b]	Moderate	40	1.1	1.67		
8	Fitchburg[c]	Municipal with significant industrial	57	15.0[e]	Moderate	15	5.4	8.00		
9	Arco Petroleum	Process waters with significant industrial	16	4.32[a]	Minimal	56	1.2	1.74	3,860	8,500
10	Rhone-Poulenc	Herbicide production waste	0.6	0.15[a]	None	87	1.4	2.00	3,860	8,500
11	Reichhold Chemicals	Chemical production waste	3.8	1.0[a]	Moderate	100	1.1	1.55	14,700	32,500
12	Stepan Chemicals	Surfactant production waste	0.06	0.015[a]	None	500			2,940	6,480
13	Republic Steel	Coke process wastes	3.6	0.95[a]	Minimal	58–116	1.6–3.1	2.3–4.6	30,800	68,000
14	Leroy	Municipal	3.8	1.0[a]	Extensive	12			5,440	12,000
15	Manchester	Water supply	150	40.0[a]	Moderate	14			5,440[d]	12,000[d]
16	Passaic Valley[e]	Water supply	8.3	2.2[a]	Extensive	8			1,090	2,400
17	Colorado Springs	Secondary effluent	7.6	2.0[a]	Extensive	17	3.1	4.5	816	1,800
18	Hercules	Chemical production	12	3.25[a]	Moderate	48	4.5	6.6	15,200	33,600
19	Industrial Sugar	Decolorizing sugar wastes	11	3.0[a]	Minimal	1080			5,440	12,000
20	Hopewell	Water supply			Moderate	7.5	1.4	2.0	None	None
21	Davenport	Water supply	110	30.0[a]	Moderate	20	1.4	2.0	None	None
22	Spreckles Sugar	Sugar thick juice			None				68,000	15,000

Source: Gumerman, Culp, and Hansen.[25]
[a] Maximum flow.
[b] Average flow.
[c] Facility under construction.
[d] Fluidized bed.
[e] GAC test facility.

Table 6. (cont'd.)

Carbon Loss (percent)	Fuel	Fuel Use		Activated Carbon Use			Cost				
		Btu/kg carbon	Btu/lb carbon	kg/ML	lb/mil gal	kg (lb)organic/ kg (lb) carbon	Capital dollars × 10³ (year)	dollars/kg/d of capacity	dollars/lb/d of capacity	dollars ML	O&M dollars/mil gal (year)
8	Gas	6,390	2,900	25	207	0.38 (0.84)	849 (1969)	311.95	141.50	9.53	30.07 (1979)
5	Propane	8,470	3,840			0.33 (0.73)	1569 (1976)	900.77	408.59		N/A[b] (1979)
10	Liquid petroleum gas[a]	6,060	2,750	30	250	0.33 (0.73)	3880 (N/A)	712.81	323.33	13.21	50.00 (1979)
9							N/A				N/A
7.5				169	1,410	0.50 (1.10)	4359 (1974)	331.37	150.31		N/A
6	Natural gas	12,300	5,600			1.70 (3.74)	3307 (1972)	607.54	275.58	23.91	90.49 (1979)
5							N/A				N/A
5	Natural gas			120	1,000	0.26 (0.57)	1000 (1971)	259.37	117.65	129.46	490.00 (1973)
8.8	Natural gas	14,300	6,500	5,150	43,000	0.38 (0.84)	300 (1969)	78.68	35.69	84.28	319.00 (1973)
5	Natural gas	13,700	6,200	9,470	79,000	0.26 (0.57)	N/A				N/A
6	Liquid petroleum gas	13,200	6,000	59,900	500,000		225 (1973)	76.54	34.72	6,605.02	25,000 (1978)
8	Natural gas			4,190	35,000		N/A	459.28	208.33		N/A
	Fuel oil						2500 (1975)				N/A
10	Fuel oil	1.5	0.7				N/A				N/A
10	Electricity						N/A				N/A
7.5				19	160	0.50 (1.10)	1622 (1973)	106.42	48.27	23.67	89.59
5	Natural gas	15,400	7,000	1,140	9,500	0.44 (0.97)	N/A			388.38	1,470.00
2.5	Natural gas						N/A			7.93	30.00 (1980)
										5.28	20.00 (1980)
4.8	Natural gas	3,940	1,785				238 (1958)	34.99	15.87		N/A

Source: Gummerman, Culp, and Hansen.[25]
[a] Data not collected.
[b] Not available.

Some of the problems encountered with GAC systems in wastewater treatment include:

- inadequate GAC transfer and feed equipment
- undersized slurry and transfer lines
- failure to provide for venting air from backwash lines, which destroys filter bottoms and disrupts GAC
- failure to house or otherwise protect automatic control systems from the weather
- inadequate means for continuous, uniform feed to the furnace, causing temperature fluctuations, inconsistent reactivation efficiency, and wasted energy
- placement of furnace and auxiliary drive motors in areas with high ambient temperature (e.g., above top of furnace)
- the use of nozzles in filter and carbon contactor bottoms, which produces major failures in carbon system just as it has for many years in water filtration plants. This use of nozzles is risky because of their tendency to clog.

In addition, there have been problems with the growth of biological organisms in the activated carbon contactors and the resulting development of anaerobic conditions with the production of corrosive hydrogen sulfide. Both of these problems have been successfully circumvented by providing adequate flow through the columns or frequent backwashing. Failure to provide adequate pretreatment has caused columns to clog and mud balls to form and has necessitated more frequent backwashing.[31]

Corrosion has been a problem with some GAC systems. The furnace system, transfer piping, and storage tanks are susceptible components. In systems using multiple-hearth furnaces, frequent replacement of the rabble arms and teeth and replacement of the hearths every few years are often required. At one installation, titanium- or ceramic-coated rabble teeth were no more resistant to corrosion than stainless-steel rabble teeth. In one case, the corrosion problem in the furnace was solved by eliminating the use of auxiliary steam during reactivation. In another, corrosion was linked to fluctuating temperatures in the hearths caused by irregular feed to the furnace and frequent startup and shutdown. In several industrial applications, the wastewater itself has been highly corrosive. In these cases, the contactors were subject to corrosion and the linings in the columns were replaced on a regular basis. Water supplies would not be expected to be so corrosive.

By properly applying the best current engineering design knowledge and practices for GAC systems, these rather serious problems might be avoided.

European Experience

Granular activated carbon has long been used in European water treatment plants.

French Experience

In France, the goal for the treatment of river waters is to produce drinking water with the minimum amount of organic content. Four categories of organic material in drinking water have been defined as follows:[32]

1. organic materials with unpleasant taste
2. organic materials without tastes, but which acquire a taste through treatment (primarily chlorination)
3. organic materials without taste and unaffected by treatment
4. organic materials that interfere with treatment (for example, by causing foaming)

For many years, because the pretreated raw waters were of adequate quality, this type of categorization was the sole one used in classifying waters for water treatment. Its principal focus was on taste and odor and, in particular, on taste as related to chlorine treatment. In France, as with many other European countries, the taste of chlorine is considered to be unacceptable. Because of this aversion to the taste of chlorine, French practice has been to use ozone. Ozone was used alone from the beginning of the century and ozone has been combined with activated carbon for the last 20 years.

Because of a recent decline in raw water quality, a new classification for organics in drinking water has been superimposed on the first one. The new classification not only is linked to the organoleptic quality of treated water but takes into consideration possible long-term effects of some chemical compounds. The new classification for organic material in drinking water is as follows:

1. organic materials of natural origin that are not transformed by treatment (especially by chlorination)
2. organic materials of natural origin that are transformed by chlorination
3. organic materials that are of synthetic origin

Eliminating the materials in the second category requires treatment without chlorination, and eliminating materials in the third category is accomplished by other treatment approaches.

Based on these new classification treatment concepts, activated carbon has evolved. In this role, carbon is not only an efficient adsorption medium but also the site for biological degradation, thereby making biochemistry a factor in water treatment technology.

Prior to 1970, activated carbon in the form of powdered activated carbon was used in France to eliminate the taste of natural organics in raw waters. These tastes were often associated with the presence of algae in rivers and in some cases were exacerbated by the action of synthetic organics, especially those produced through chlorination. However, as water quality has continued to deteriorate, there has been a constant increase in the level of chlorination accompanying the rise of dissolved ammonia in raw waters. Chlorination has created taste problems and powdered activated carbon has become less effective. Therefore, at the end of the 1960s, French engineers responsible for water treatment considered the possibility of granular activated carbon. Initially, this modification was the simple replacement of sand by granular activated carbon in the filter bed.

Table 7 lists the utilities that have been using powdered activated carbon as the carbon treatment process, and Table 8 lists those plants that are using GAC by replacing sand in the filter shell. In these plants, good results have been obtained for the removal of the chlorine taste through simple filtration by activated carbon. A number of plants have applied or are still applying the technique of breakpoint chlorination followed by flocculation, sedimentation, and activated carbon filtration (Table 9).

Disinfection is usually accomplished by ozonation, and network protection is accomplished through the addition of sodium hypochlorite or gaseous chlorine or chlorine dioxide to the treated waters. Carbon is regenerated every year or every two years, depending upon the individual utility. Some carbon beds have been regenerated up to five times without difficulty. Regeneration (reactivation) criteria are most often based on the taste threshold test. Filter types are usually the open type made of concrete in rectangular shapes with a service area of about 220 m² (263 sq yds). Carbon depth usually varies from 0.7 to 1.0 m (2.3–3.3 ft) and empty bed contact time varies from 6 to 20 min, respectively.

As water quality has continued to deteriorate, a new technique has been developed that combines step ozonation with sand and activated carbon double filtration. The second filtration with carbon is always preceded by ozonation, which efficiently reduces or eliminates organic substances. This issue will be discussed in more detail later. This concept is being considered by the largest treatment plants in France. Table 9 contains a listing of those plants that are practicing step ozonation before a second filtration with granular activated carbon. Ozonation in combination with activated carbon constitutes what is frequently called biological activated carbon (BAC). Ammonia is biologically eliminated in the sand filter for the most part, while the activated carbon filter through the BAC process completes the reduction of organics.

Step ozonation may be carried out in many different ways, but the common feature is that at least one ozonation step is used before the second filtration with activated carbon.

Table 7. Principal Drinking Water Plants in France Using Powdered Activated Carbon (Output > 50×10^3 m³/day)[a]

Plant	Output[a] m³/day	Raw Water Origin	District Supplied	Treatment	Date of PAC Going Into Service	Range of Treatment Rate (mg/L)	PAC Function	Storage and Supply	Carbon Origin
Choisy-le-Roi	800,000	Seine River	Paris—southern suburbs	Chlorination; aluminum polychloride coagulation; powdered activated carbon treatment; flocculation-sedimentation; rapid sand filtration, part on sand, part on activated carbon; ozonation; addition of Cl_2	1964	0 to 30	Tastes	2 silos (90 T each); weighing systems	Coconut
Neuilly-sur-Marne	600,000	Marne River	Paris—southern suburbs	Chlorination; aluminum polychloride coagulation; powdered activated carbon treatment; flocculation-sedimentation; rapid sand filtration; ozonation; addition of Cl_2	1968	0 to 30	Tastes	1 silo (70 T) (plus containers); screw conveyor; proportioning	Coconut
Orly	300,000	Seine River	Paris—southern suburbs	Chlorination; $Al_2(SO_4)_3$ plus activated silica coagulation; powdered activated carbon treatment; flocculation-sedimentation; rapid sand filtration; ozonation; addition of ClO_2	1970	0 to 20	Tastes	3 silos; 2 gravimetric batchers	Coconut and peat
Mery-sur-Oise	200,000	Oise River	Paris—northern suburbs	Present treating line; chlorination; aluminum polychloride coagulation; powdered activated carbon treatment; flocculation-sedimentation; rapid sand filtration, part on sand, part on activated carbon; ozonation; addition of ClO_2	1964	0 to 30	Tastes; organic matter removal	Containers; proportioning; screw conveyors	Coconut

Source: Schulhof.[32]

[a]Multiply m³/day by 0.183 to obtain gal/min.

Table 7. (cont'd.)

Plant	Output[a] m³/day	Raw Water Origin	District Supplied	Treatment	Date of PAC Going Into Service	Range of Treatment Rate (mg/L)	PAC Function	Storage and Supply	Carbon Origin
Croissy	120,000	Seine River	Outer western suburbs	Coagulation; flocculation (aluminum sulfate plus activated silica); powdered activated carbon; sedimentation with powdered activated carbon; aerated sand filtration; ground level replenishing through sand pits	1958	10	Detergent removal (avoiding foams in sand pits)	Containers; proportioning	Coconut and peat
Aubergenville	120,000	Seine River	Outer western suburbs	Coagulation; flocculation (aluminum sulfate plus activated silica); powdered activated carbon; sedimentation with powdered activated carbon; nitrification (bacteria beds); sand filtration; ozonation; addition of Cl_2	1961	Varies widely	Tastes (in connection with algae and bacteria)	Containers; proportioning	Coconut and peat
Morsang (first section)	75,000	Seine River	Outer southern suburbs	Prechlorination; coagulation (aluminum sulfate plus activated silica); powdered activated carbon; flocculation-sedimentation with powdered activated carbon; sand filtration; ozonation; addition of Cl_2	1969	15	Tastes; organics removal	1 concrete silo (60 m³)	Peat and anthracite
Suresnespont-Valerien	50,000	Seine River	Paris northwestern suburbs	Chlorination; powdered activated carbon; coagulation; flocculation-sedimentation; rapid sand filtration; addition of ClO_2	1959	40	Tastes	Containers; proportioning	Peat and anthracite

Source: Schulhof.[32]
[a]Multiply m³/day by 0.183 to obtain gal/min.

Table 8. Principal Drinking Water Plants in France Using Granular Activated Carbon for First Filtration (Output $>50 \times 10^3$ m³/day)

Plant	Capacity[a] (m³/day)	Raw Water Origin	District Supplied	Treatment	Date of GAC Going Into Service	GAC Function	Number of GAC Filters	Filter[b] Volume per Unit (m³)	Average Empty Bed Contact Time (min)	Regeneration Period (years)	Carbon Type
Mery-sur-Oise	100,000	Oise River	Paris northern suburbs	Chlorination; aluminum polychloride coagulation; flocculation-sedimentation; GAC filtration; ozonation; chlorination	1970	Tastes; organic matter	6	108	15	12	1 filter Picactif; 2 filters Picafil; 2 filters Picaflo G; 1 filter F300
Viry-Chatillon	100,000	Seine River upstream of Paris	Paris southern suburbs	Prechlorination; aluminum polychloride coagulation; flocculation-sedimentation; GAC filtration; ClO₂ disinfection	1973	Tastes; organic matter	15	9 at 40; 6 at 32	10	30	F300 E; Row 0.8 supra
Saint-Charles	100,000	Moselle River	Nancy	Prechlorination; coagulation plus flocculation-sedimentation; GAC filtration; ClO₂ disinfection	1974	Tastes; organic matter	22	48	15	36	NCY 103 at F300
Vigneux-sur-Seine	50,000	Seine River upstream of Paris	Outer western suburbs	Prechlorination; aluminum; sulfate plus activated silica coagulation; flocculation-sedimentation; GAC filtration; ClO₂ disinfection	1970	Tastes; organic matter	10	23	6	12	F300; F500; Row 0.8 supra; BS12

[a]Multiply m³/day by 0.183 to obtain gal/min.
[b]Multiply m³ by 35.314 to obtain ft³.

Table 8.   (cont'd.)

Plant	Capacity[a] (m³/day)	Raw Water Origin	District Supplied	Treatment	Date of GAC Going Into Service	GAC Function	Number of GAC Filters	Filter[b] Volume per Unit (m³)	Average Empty Bed Contact Time (min)	Regeneration Period (years)	Carbon Type
Houille	50,000	Houille River	Dunkerque	Prechlorination; coagulation; flocculation (Fe-ClO$_2$ plus starch); sedimentation, in part; flotation, in part; GAC filtration basins	1973	Organic matter	7	50	10	6	Row 0.8 supra
Choisy-le-Roi	30,000	Seine River	Paris southern suburbs	Chlorination; aluminum polychloride coagulation; flocculation-sedimentation; GAC rapid filtration; ozonation; chlorination	1978	Tastes	2	117	15	Not yet regenerated	1 filter Picaflo B 1 filter F300
Annet-sur-Marne	25,000	Marne River	Paris eastern suburbs	Aluminum polychloride coagulation; flocculation-sedimentation; activated carbon filtration; ozonation; chlorination	1971	Tastes; organic matter	2	45	13	36	1 filter Picactif 1 filter Picaflo

aMultiply m³/day by 0.183 to obtain gal/min.
bMultiply m³ by 35.314 to obtain ft³.

Table 9. Principal Drinking Water Plants Using Step-Ozonation Before a Second Filtration on Granular Activated Carbon

Plant	Capacity[a] (m³/day)	Raw Water Origin	District Supplied	Treatment	Date of GAC Going Into Service	GAC Function	Number of GAC Filters	Filter[b] Volume per Unit (m³)	Average Empty Bed Contact Time (min)	Regeneration Period (years)	Carbon Type
Mery-sur-Oise	300,000	Oise River	Northern suburbs	First ozonation; raw water storage; coagulation by aluminum chloride; flocculation-sedimentation; sand filtration; second ozonation; GAC filtration; chlorine disinfection; dechlorination; ClO₂	1979	Organics elimination; chlorine stability and NH₄ elimination	12	108	13	Not yet regenerated	2 Picactif 2 Picafil 4 Picaflo G 2 F300
Morsang (second section)	75,000	Seine River upstream of Paris	Remote southern suburbs	Prechlorination; coagulation; flocculation-sedimentation; sand filtration; ozonation (half the plant); GAC filtration; final disinfection chlorination	1975	Organics elimination	4	90	10	3	F400
Rouen-la Chapelle	50,000	ground-water	Rouen southern suburbs	First ozonation; sand filtration; activated carbon filtration; second ozonation; chlorination	1975	Micropollution and NH₄ elimination	6	75	13	1 filter regenerated after 4 years testing	5 Picactif 1 Picafil

Source: Schulhof.[32]
[a]Multiply m³/day by 0.183 to obtain gal/min.
[b]Multiply m³ by 35.314 to obtain ft³.

Experience in The Netherlands

The annual production of drinking water in The Netherlands amounts to one billion m³ (1.3 × 10⁹ yd³).[33] The main sources of water are the Rhine and Meuse Rivers, the Yssel Lake, and the Haringvliet. The Rhine is polluted by many contaminants, while the other sources are somewhat less polluted. Given the problems of quality in the available sources, Dutch waterworks policy requires that at least one month's reserve supply of treated water be available. Reserve water supplies are collected in storage basins and by dune infiltration. Selective intake of raw water prevents serious pollution from reaching the treatment plants. Collection of reserve supplies substantially improves and equalizes water quality, but pollution is so great that treatment with activated carbon is frequently required. Powdered activated carbon is widely used, but many waterworks are beginning to change over to granular activated carbon treatment.

Belgian Experience

Belgian water utilities rely on groundwater for about 80% of their water supply.[34] Powdered activated carbon is used in the Tailfer Plant of the Brussels Intercommunal Waterboard (CFIBE) to improve color and to remove taste and odor in conjunction with ClO_2 to reduce trihalomethane formation potential. The low level of trihalomethanes found in the water supply in Belgium does not justify the introduction of GAC in most plants, although the technique is being considered seriously. Several pilot plants have been operated to investigate the use of GAC for abatement of TOC and THM and as biologically active filters.

Swiss Experience

The Lake Waterworks in Switzerland uses granular activated carbon filtration, despite the relatively high quality of their source water.[35] This is due to the proximity of oil pipelines to drinking water storage basins and the occurrence of two phenol spills in St. Gallen and Zurich drinking water supplies. When the oil pipeline from Genoa to Ingolstadt was laid adjacent to Lake Constance, the possibility of oil leakage was considered. As a result, ozone or GAC treatment facilities were installed in all of the waterworks in the critical area. Thus, water contaminated by slight amounts of oil could be purified for potable use.

Phenol spills that contaminated the water supplies of the cities of St. Gallen in 1957 and Zurich in 1967 demonstrated the need for GAC filtration. On both occasions drinking water supplies were highly contaminated and taste and odor were adversely affected. These two events led to the installation of a GAC filtration system at the St. Gallen waterworks by

Lake Constance. The phenol spill in Zurich had consequences so severe that it provided sufficient impetus for the installation of a GAC filtration step at the Lengg and Moos treatment plants.

UK Experience

In the United Kingdom, water suppliers are legally required to supply "wholesome water," which is defined as water that is generally palatable and safe.[36] Approximately 100 potable water treatment works in the country have the capability of using activated carbon. In most cases activated carbon is used for controlling taste and odor problems, particularly earthy, musty tastes and odors. These taste and odor problems are seasonal and often short-lived, so that most water utilities find it more economical to use powdered rather than granular activated carbon. The UK has four water treatment plants at present that employ granular activated carbon, three of which use the carbon after rapid filters and one in which the carbon has partially replaced the media in the rapid filters. One of the treatment plants employs granular activated carbon to remove "chemical" taste and odor and the other three are for the control of earthy, musty taste and odor. Only one of the treatment plants has regeneration facilities onsite.

German Experience

Since 1975, research into the use of granular activated carbon has been conducted in Germany at an accelerated pace.[37] The German Ministry of Research and Development has subsidized investigations designed to study the optimization of carbon use and regeneration, and to examine biological treatment with activated carbon filters and on macroreticular resins for humic acid removal. Many of these studies are conducted onsite at water treatment plants focusing on the practical aspects of activated carbon use. In the following section, the experience of using GAC in Germany is discussed.

Langenau Water Works (Langenau, Federal Republic of Germany). The Langenau water treatment plant is on the Danube River with a capacity of 2.3 m³/sec (52 mgd). In this plant, after ozonation, water goes on to a filter with mixed media above granular activated carbon filters. The rapid sand filters operate at a relatively high velocity of 15 m/hr (6 gal/min/ft²). The upper layer consists of 1.2 m (3.9 ft) of sand with a diameter of 1.5 to 2.5 mm (0.06–0.10 in.), and the lower layer is 1.2 m (3.9 ft) of sand with a diameter of 0.6 to 1 mm (0.02–0.03 in.). Below the sand is a layer of gravel on a filter bottom. The filters are backwashed every 4–6 days. Below the filter bottom is a second chamber containing 1.5 m (4.9 ft) of GAC and below that 0.4 m (1.3 ft) of sand. In Langenau, the current bed life of GAC is approximately 440 days.

Stadwerke Dusseldorf AG (Dusseldorf, Federal Republic of Germany).
Dusseldorf has three water treatment plants. One is Staad, the second is
Flehe, and the third is Holthausen. The Flehe plant was built in 1871 and
consists only of sand bank filtration with wells driven in the banks of the
Rhine River. There is an impervious zone of clay laid along the river bank,
so the water is forced to come down through the river bed, over through the
sand, and up to the wells, providing a long flow path. In 1967, the plant was
enlarged to its present capacity of 29.4 ft³/sec (19 mgd) to include ozone
oxidation of manganese and granular activated carbon adsorption. Probably the most important aspect of this treatment plant was the pressure filters
that were built outside the treatment plant. Ozonation is applied at a 3
mg/L dosage for a 5-min contact time and then 30 min of storage is used to
allow the ozone to dissipate.

Once the water is pumped through the double filters, it goes directly to
the user without any repumping. The upper filter has 1 m (3.3 ft) of semiac-
tivated carbon with a gradation between anthracite and granular activated
carbon, with 30 cm (1 ft) of gravel below the carbon. The lower compartment consists of 2 m (6.6 ft) of granular activated carbon supported by 30
cm (1 ft) of gravel. The upper filter is backwashed every day at peak loads
and every other day at normal loads: the granular activated carbon is backwashed about one-third of the number of times that the upper filter is
backwashed. The filters have 300 m³ (10,600 ft³) of GAC to treat the 3000-
m³/hr (8.4-mgd) flow. The filtration rate is 22 m/hr (8.8 gpm/ft²). The final
treatment step is the addition of chlorine dioxide.

The reactivation furnace is located at the Holthausen works. This plant is
a 10,000-m³/hr (63.5-mgd) plant (currently being doubled in capacity)
employing bank filtration, ozonation for manganese control (3 mg/L), 5-
min contact time, and pressure filter-adsorbers with 1.6 m (5.2 ft) of filter
media in the upper stage and 2.5 m (8.2 ft) of GAC in the lower stage. The
flow rate is 12–15 m/hr (4.8 to 6 gpm/ft²). Final disinfection is with 0.2 to
0.3 mg/L ClO$_2$ generated from chlorine and sodium chlorite with 10%
excess chlorine.

The Staad waterworks contains 600 m³ (20,600 ft³) of GAC, Flehe contains 300 m³ (10,600 ft³) of GAC, and Holthausen contains 1200 m³ (42,400
ft³) of GAC. Therefore, Holthausen was the choice for the location of a
reactivation furnace. All three plants have about 110 metric tons (123 tons)
of GAC. The Dusseldorf waterworks rents trucks to bring GAC from the
other two plants to the Holthausen plant.

GAC is brought directly from the filters under pressure at the Holthausen
plant and put in storage bins, where some separation from the water occurs.
A 3:1 water-to-GAC ratio is maintained for moving the GAC throughout
the plant. The water moves through plastic pipes at a rate of 1.3 m/sec (4.3
ft/sec). Following dewatering, the GAC goes to the two-bed fluidized bed
furnace. The furnace's inside dimensions are 3 ft (0.9 m) in diameter by 6.6

ft (2.0 m) high. The outside diameter is 4.3 ft (1.3 m). The unit is a 250-kg/hr (550-lb/hr) furnace and is fully automated. There is room in the building for two additional furnaces of equal size and one additional larger offgas-handling facility.

The rate of flow of GAC through the furnace is controlled by temperature, so that if the temperature goes below an optimum level, the rate of GAC feed is slowed down. The fine materials separated by the cyclone separator are collected and sent to a wastewater sewer. By law, the gaseous effluent discharge must be 400°C (752°F), so there is no cooling of the offgases after they are burned. The intent is to operate this furnace 24 hours a day, 7 days a week, and have the entire operation controlled by computer, even though there will be staff onsite during the day. At these rates, it would take about 170 days to reactivate all the GAC in the three plants. When one storage bin is full of GAC, it takes 3 days for the furnace to reactivate it. Exhaustion occurs in approximately 95 days.

Wuppertal Stadtwerke AG (Woppertal, Federal Republic of Germany). Water is aerated and ozone is added to the aerated water and passed over a dry (trickling) filter for manganese removal and then down through an activated carbon filter prior to chlorination and distribution.

The filters contain 1.5 m (4.9 ft) of granular activated carbon (Chemviron F300 GAC). Also used is a new support material from Chemviron called BPL 4 × 10, which is a heavy GAC material that is not abrasive on the furnace if some is accidentally taken to the furnace. More water is used for GAC movement than in Dusseldorf. Water use is 1000 m³ (35,000 ft³) for every 20 metric tons (22.4 tons) of GAC (23:1 on a volume basis) to avoid attrition. The filtration rate is 14 m/hr (5.6 gpm/ft²).

REFERENCES

1. Clark, R. M. "The Safe Drinking Water Act: Its Implication for Planning," *Municipal Water System – The Challenge for Urban Resource Management*, D. Holtz and S. Sabastian, Eds. (Bloomington, IN: Indiana University Press, 1978).
2. Middleton, F. M., and A. A. Rosen. "Organic Contaminants Affecting the Quality of Water," *Public Health Rep.* 71(11):1125–1133 (1965).
3. Public Health Service Drinking Water Standards, PHS Publication 956, U.S. Government Printing Office (1962).
4. Bellar, T. A., J. J. Lichtenberg, and R. C. Kroner. "The Occurrence of Organohalides in Chlorinated Drinking Water," *J. Am. Water Works Assoc.* 66:703–706 (1974).
5. Rook, J. J. "Formation of Haloforms During Chlorination of Natural Waters," *Water Treatment Exam.* 23(2):234–243 (1974).
6. Coleman, W. E., et al. "The Occurrence of Volatile Organics in Five Drinking Water Supplies Using Gas Chromatography/Mass Spectrom-

etry," in *Identification and Analysis of Organic Pollutants in Water*, L. H. Keith, Ed. (Ann Arbor, MI: Ann Arbor Science Publishers, Inc., 1976), pp. 305–347.

7. Lingg, R. D., et al. "Quantitative Analysis of Volatile Organic Compounds by GC-MS," *J. Am. Water Works Assoc.* 69(11):605–612 (1977).

8. "Report on the Carcinogenesis Bioassay of Chloroform," National Cancer Institute, 1976.

9. Page, T., R. H. Harris, and S. S. Epstein. "Drinking Water and Cancer Mortality in Louisiana," *Science*, 193:55 (1976).

10. "New Orleans Area Water Supply Study," Surveillance and Analysis Division, EPA Region VI, EPA Report 906/9-75f-003, Dallas, TX (1975).

11. "Control of Trihalomethanes in Drinking Water: Final Rule," National Interim Primary Drinking Water Regulations, *Federal Register* 44(231): 68624–68705 (1979).

12. Hueper, W. C., and W. D. Conway. *Chemical Carcinogenesis and Cancer* (Springfield, IL: Charles C. Thomas Publisher, 1964).

13. Symons, J. M., et al. "National Organics Reconnaissance Survey for Halogenated Organics," *J. Am. Water Works Assoc.* 67:634 (1975).

14. Page, T., R. H. Harris, and J. Bruser. "Removal of Carcinogens from Drinking Water: A Cost-Benefit Analysis," Social Services Working Paper 230, Division of Humanities and Social Sciences, California Institute of Technology, Pasadena, California (January, 1979).

15. Westrick, J. J., W. Mello, and R. Thomas. "The Groundwater Supply Survey," *J. Am. Water Works Assoc.* 76(5):52–59 (1984).

16. Dyksen, J. E. "Treatment Techniques for Removing Organic Compounds from Groundwater Supplies," *Ind. Water Eng.* 19(4):16–21.

17. Lykins, B. W. Jr., and J. A. Baier. "Removal of Agricultural Contaminants from Groundwater," American Water Works Association Annual Conference, June 23–27, 1985.

18. Litwin, Y. J., N. H. Hantzsche, and N. George. "Groundwater Contamination by Pesticides – A California Assessment," submitted to the State Water Resources Control Board, Sacramento, CA, by Ramlit Associates, Inc., Berkeley, CA.

19. Safe Drinking Water Act, Public Law 93-523 (1974).

20. Symons, J. M., et al. "Treatment Techniques for Controlling Trihalomethanes in Drinking Water," Municipal Environmental Research Laboratory, Office of Research and Development, U.S. EPA Report 600/2-81-156 (1981).

21. Symons, J. M. "A History of the Attempted Federal Regulation Requiring GAC Adsorption for Water Treatment," *J. Am. Water Works Assoc.* 76(8):34–43 (1984).

22. Minutes of National Drinking Water Advisory Council Meeting, August 22–23, 1978, and letter from Council Chairman C. C. Johnson, Jr., to Administrator Costle, September 29, 1978, NOWAC, Office of Drinking Water, U.S. EPA, Washington, DC.

23. The Safe Drinking Water Act Amendments, Public Law 99-339 (1986).
24. Gumerman, R. C., R. L. Culp, and S. P. Hansen. "Estimating Water Treatment Costs: Vol. 1. Summary," Municipal Environmental Research Laboratory, U.S. EPA Report 600/2-79-162a (1979).
25. Gumerman, R. C., R. L. Culp, and S. P. Hansen. "Estimating Water Treatment Costs: Vol. 2. Cost Curves Applicable to 1 to 200 mgd Treatment Plants," Municipal Environmental Research Laboratory, U.S. EPA Report 600/2-79-162b (1979).
26. Gumerman, R. C., R. L. Culp, and S. P. Hansen. "Estimating Water Treatment Costs: Vol. 3. Cost Curves Applicable to 2600 mgd to 1 mgd Municipal Environmental Research Laboratory Treatment Plants," U.S. EPA Report 600/2-79-162c (1979).
27. Gumerman, R. C., R. L. Culp, and S. P. Hansen. "Estimating Water Treatment Costs: Vol. 4. Computer User's Manual for Retrieving and Updating Cost Data," Municipal Environmental Research Laboratory U.S. EPA Report 600/2-79-162d (1979).
28. Culp, R. L., and R. M. Clark. "Granular Activated Carbon Installations," *J. Am. Water Works Assoc.* 75(8):398-405 (1983).
29. Lanonette, K. H. "Treatment of Phenolic Wastes," *Chem. Eng.* (Deskbook Issue, October 1977), p. 104.
30. Culp, R. L. "GAC Water Treatment Systems," *Public Works* (February 1980), pp. 83-87.
31. Culp, R. L., G. Wesner, and G. L. Culp. *Handbook of Advanced Wastewater Treatment* (New York: Van Nostrand Reinhold Co., 1978).
32. Schulhof, P. "An Evolutionary Approach to Activated Carbon Treatment in France," *J. Environ. Pathol. Toxicol. Oncol.* (718):55-75 (1987).
33. Meijers, A. P., J. J. Rook, B. Schultnik, J. Smenck, G. M. M. Laan, J. Vander Poels, and L. M. Cees. "Objectives and Procedures for GAC Treatment in The Netherlands," *J. Environ. Pathol. Toxicol. Oncol.* 7(718):55-75 (1987).
34. Masschelein, W. J. "Practical Applications of Adsorption Techniques in Drinking Water—Belgium Experiences," *J. Environ. Pathol. Toxicol. Oncol.* 7(718):55-75 (1987).
35. Schalekamp, M. "The Use of GAC Filtration to Ensure Quality in Drinking Water Sources from Surface Sources—Swiss Experience," *J. Environ. Pathol. Toxicol. Oncol.* 7(718):55-75 (1987).
36. Goodall, J. B., and R. A. Hyde. "Current United Kingdom Practice in the Use of Granular Activated Carbon in Drinking Water Treatment," *J. Environ. Pathol. Toxicol. Oncol.* 7(718):55-75 (1987).
37. Symons, J. M. "Trip Report for European Travel with Special Emphasis on Reactivation of Granular Activated Carbon," Water Supply Branch, Municipal Environmental Research Laboratory, U.S. EPA (April 1977).

CHAPTER 2

GAC Process Design Considerations

INTRODUCTION

GAC has been designated as a baseline technology for the removal of organic contaminants from drinking water. Although activated carbon has been used for many years in both drinking water and wastewater treatment and is often referred to as a single process, it is really a series of unit process modules.[1] These modules and their associated design and operational considerations are discussed in detail in the following section.

DEFINITION OF PROCESS MODULES

Activated carbon adsorption is based on the ability of specially prepared carbon to remove certain chemical species from a liquid solution by adsorption. In water treatment, the carbon can be either powdered activated carbon, consisting of particles at and below U.S. Sieve Series No. 50, or granular activated carbon, consisting of larger particles.[2] The adsorptive properties of the PAC and GAC are not fundamentally different, since they depend on pore size and the internal surface area of the pores, which are independent of overall particle size.

Each commercially available granular activated carbon has properties making it more suitable for certain applications than others. Besides adsorptive capacity and selectivity, these properties include the ability to withstand thermal reactivation and resistance to attrition losses during transport and handling. Table 1 lists typical commercially available car-

31

Table 1. Properties of Typical Granular Activated Carbons

	Type A	Type B	Type C	Type D
Physical Properties				
Surface area, m^3/g (BET)	600–630	950–1050	1000	1050
Apparent density, g/cm^3	0.43	0.48	0.48	0.48
Density, backwashed and drained, lb/ft^3	22	26	26	30
Real density, g/cm^3	—	2.1	2.1	2.1
Particle density, g/cm^3	2.0	1.3–1.4	1.4	0.92
Effective size, mm	1.4–1.5	0.8–0.9	0.85–1.05	0.89
Uniformity coefficient	0.8–0.9	1.9	1.8	1.44
Pore volume, cm^3/g	1.7	0.85	0.85	0.60
Mean particle diameter, mm	1.6	1.5–1.7	1.5–1.7	1.2
Specifications				
Sieve size, U.S. std. series				
Larger than No. 8 max. percentage	8	8	8	—
Larger than No. 12 max. percentage	—	—	—	5
Smaller than No. 30 max. percentage	5	5	5	—
Smaller than No. 40 max. percentage	—	—	—	5
Iodine No.	650	900	950	1000
Abrasion No., minimum	—	70	70	85
Ash, percentage	—	8	7.5	0.5
Moisture as packed, max. percentage	—	2	2	1

— = Data not available.

bons.[2] Detailed discussions of properties and comparisons of different carbons and adsorption mechanisms are available in published literature.[1,3]

The practical application of granular activated carbon in water and wastewater treatment relies on the reuse of most of the carbon. During use, the carbon gradually becomes saturated with the species being adsorbed (adsorbate) so that it eventually loses its capacity to adsorb more material. Without a method to restore its capacity, GAC would be prohibitively expensive except for those applications where carbon usage was small.

The most common method for restoring capacity is thermal regeneration or reactivation. In this process, spent carbon is heated in a furnace to a high temperature and adsorbed materials are driven off and burned. The process is controlled to avoid burning the carbon and minimize carbon losses due to both oxidation and abrasion. After reactivation, the carbon is then returned to the adsorbers for further service.

A conceptual diagram of a granular activated carbon system is shown in Figure 1, consisting of the following major process modules: carbon adsorption, reactivation, and carbon storage and transfer.

The heart of the system is the adsorbers, which contain the granular activated carbon. An influent water stream passes through the adsorbers where contact with the carbon removes contaminants. Backwash and surface wash equipment, as well as influent pumping, can be considered as components of this subsystem.

Spent carbon is removed from the adsorbers and transferred to the reacti-

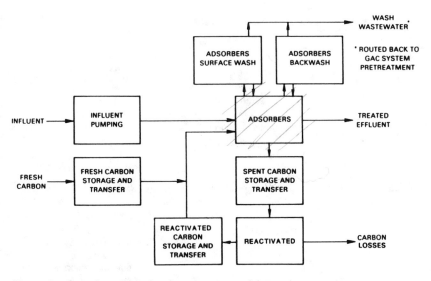

Figure 1. Granular activated carbon process modules.

vation subsystem for restoration of its adsorptive capacity. Reactivated carbon is returned to the adsorbers, and fresh makeup carbon is added to compensate for losses during transfer and reactivation. The reactivation subsystem consists of the furnace, fuel system, air pollution control equipment for reactivator stack gases, and carbon furnace feed and discharge equipment.

The carbon storage and transfer subsystem consists of equipment for transfer of carbon between adsorbers and reactivation, as well as fresh carbon unloading, transfer, and storage. Slurrying the carbon with water is the most common transfer method. Storage vessels are used for fresh carbon, spent carbon between adsorbers and reactivation equipment, and reactivated carbon between reactivation equipment and adsorbers.

Each of these major subsystems consists of various groupings of equipment items defined as process modules. Key design parameters are shown in Table 2. Values shown in the table are typical and may vary with specific applications. Some of these design parameters are discussed in the following paragraphs.

CARBON SELECTION

The first consideration in the design of any activated carbon system is carbon selection. Selection depends on the ability of a given carbon to remove the contaminants of concern and to meet other system requirements related to pressure drop (head loss), carbon transport, and reactivation.

A number of granular activated carbons are commercially available, and

Table 2. Typical Design Parameter Ranges for GAC Systems

Subsystem	Parameter	Typical Values
Adsorbers	Carbon type	Varies with application
	Contact time	10–20 min
	Breakthrough characteristics	Varies with influent properties and carbon type
	Hydraulic loading	2–10 gpm/ft²
	Backwash frequency and rate	15–20 gpm/ft² for periods of 10–20 min several times a day
Reactivation	Carbon throughput rate	Up to 50,000 lb/hr/unit
	Operating temperature	1200–1900°F
	Residence time	10 min–4 hrs, depending on reactivation method and the nature of the materials adsorbed on the carbon
Carbon storage and transfer	Slurry flow rate	Varies with discharge interval selected by designer
	Reactivation frequency	Intermittent and variable; usually several weeks to months
	Carbon losses	5–10% of reactivation quantity
	Transport slurry composition	10 lb water per lb of carbon

the type most suited for a given application is usually determined by laboratory and pilot testing based on isotherms.[3] Although mathematical models have been developed to predict adsorption behavior in carbon systems, the complexity of most systems still requires experimental design data.[4,5] Models can be useful for preliminary design and cost analyses where an estimate of performance may be required.

The two most commonly available granular activated carbon particle sizes are 8 × 30 and 12 × 40 mesh. These designations are based on 95% of the carbon passing through the mesh size.

CONTACT TIME AND BREAKTHROUGH

Two important variables in carbon system design are contact time and breakthrough characteristics. These variables, along with flow quantity, determine carbon bed and adsorber volume. Carbon exhaustion or usage rate, which is related to the breakthrough characteristic, will determine the size of the reactivation system. Size of the adsorbers and reactivation system and scheduling of carbon transfer between these two systems determine carbon transfer and storage requirements.

Empty bed contact time (EBCT) is the carbon bed volume divided by the superficial flow rate of the fluid stream through the adsorber. Adsorber volume depends on bed volume and how much freeboard or excess vessel capacity over bed volume is provided. Freeboard may range up to about 50% for fixed bed and expanded bed systems. If bed expansion is not required, even with backwashing to remove collected solids, a freeboard of 20-30% may be adequate. Virtually no freeboard is required for upflow pulsed beds.

Contact time can be varied by changing bed depth at constant flow, or changing flow at constant bed depth. The effect of contact time on performance is to alter the time to breakthrough which is defined as the point where the solute concentration in the adsorber effluent exceeds the treatment objective. Shorter relative contact times result in earlier breakthrough; longer relative contact times delay breakthrough. There is clearly an economic tradeoff between breakthrough frequency and adsorber volume.

In addition to delaying breakthrough up to a certain point, carbon utilization improves as contact time increases. The percentage of total carbon in the bed that is exhausted at breakthrough is greater in a deeper than in a shallower bed. Beyond a certain point, however, additional adsorber volume merely acts as storage capacity for spent carbon. There is, therefore, an optimum bed depth from the perspective of adsorber cost alone. The actual choice of bed depth and corresponding adsorber volume, however, also depends on reactivation frequency. A capital cost versus operating cost tradeoff exists between increased adsorber volume with reduced reactivation frequency and reduced adsorber volume with increased reactivation frequency. The optimum bed depth and the tradeoffs will depend on individual system influent and carbon bed characteristics.

BREAKTHROUGH CHARACTERISTICS

Breakthrough is defined to occur when the effluent concentration exceeds some preset value, but breakthrough characteristics are a continuous function of effluent concentration versus total volume of water treated, or versus time. Breakthrough is described mathematically by solution of the differential equations describing adsorption. Figure 2 illustrates a typical breakthrough curve and relates it to the relative amount of the carbon bed exhausted, still active (adsorption zone), and as yet relatively free of adsorbed material. At startup, effluent emerging from the column has a very low concentration of solute. Most of the solute has been adsorbed by the carbon near the beginning of the bed. As more liquid flows through the

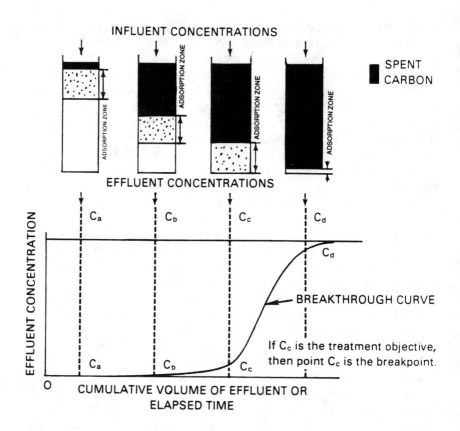

Figure 2. Passage of adsorption wave through a fixed bed and corresponding breakthrough curve.

column, the adsorption capacity of the carbon is exhausted, less solute is
adsorbed, and a gradual increase in effluent solute concentration occurs.
Finally, as the bed nears exhaustion, the effluent solute concentration
increases rapidly as it approaches the influent solute concentration.

Breakthrough depends on the characteristics of both the influent stream
and the carbon bed. Different solutes with different carbons will yield
different slopes for breakthrough curves at a given flow rate and bed volume (i.e., a given contact time). Also, the time interval from startup to the
breakpoint shown in Figure 2 will depend on the specific solute and carbon.
The two physical-chemical variables for a given system which determine
these breakthrough characteristics are the equilibrium constant and the rate
constant for adsorption.

Breakthrough curves are important to design because they define the
relationship between the physical-chemical parameters of the solvent-
solute-carbon system, including flow rate, bed size, carbon exhaustion rate,

the selection of the number of columns in a series multiadsorber system, and the treatment objective. Figure 3 illustrates typical adsorption breakthrough curves showing effect of contact time and exhaustion.

BREAKTHROUGH AND SYSTEM DESIGN

There is a relationship between the breakthrough characteristics, treatment objective, and number of adsorbers.[6] For example, consider a single adsorber (Figure 4). Three treatment objectives are illustrated and labeled A, B, and C along with their corresponding effluent concentrations and time in operation. Consider A as a baseline treatment objective. When a lower effluent concentration than A is required, the time in operation for the carbon bed is reduced from t_A to t_B. For a given contact time, the design implication is that the carbon exhaustion rate is higher, carbon dosage is higher, and reactivation must be more frequent.

For a fixed treatment objective, analysis of several breakthrough curves for an influent under study provides information needed to size the adsorber and reactivation system. The contact time selected for design will be the one which yields a reasonable adsorber volume and reactivation frequency. The choice depends on the cost tradeoff between having a large carbon bed and adsorber volume with less frequent reactivation, and a smaller adsorber volume and more frequent reactivation.

In addition to frequency of reactivation, the breakthrough curve provides information for estimating the minimum reactivation rate required. If the changeover time for emptying spent carbon and recharging reactivated carbon in an adsorber is negligible compared to total onstream time, then the minimum time during which the carbon is reactivated must be less than or equal to the length of onstream time before breakthrough. This is to ensure that reactivated carbon will be ready to fill the adsorber just as the carbon in use is exhausted. Surge storage capacity can be provided so that the actual reactivation rate can differ from the value obtained using this minimum reactivation time duration.[7]

In practice, multiple beds in series can be used. When one bed is exhausted and ready for reactivation, the next bed still has some capacity left, ensuring that the treatment objective will not be exceeded at any time. In essence, this enables a portion of the total carbon bed to be removed and reactivated while providing continuous treatment with the remaining portion. Carbon transfer to the regeneration system is more frequent, but if the total volume of the two series beds is equivalent to the volume for a single bed to accomplish the same treatment objective, the reactivation feed rate of carbon is unchanged. Multiple beds in series also provide for a combination of upflow and downflow vessels, which could increase the process efficiency.

CHOICE

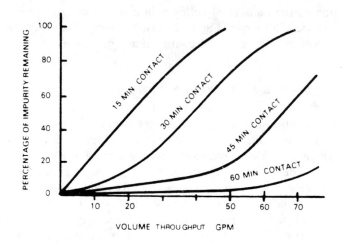

(A)

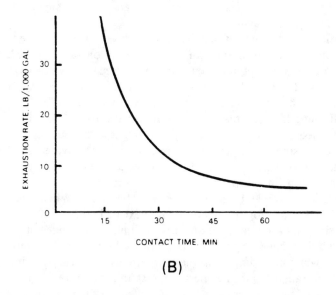

(B)

Figure 3. Typical carbon adsorption breakthrough curves showing effect of contact time (A) and corresponding exhaustion rate curve (B).

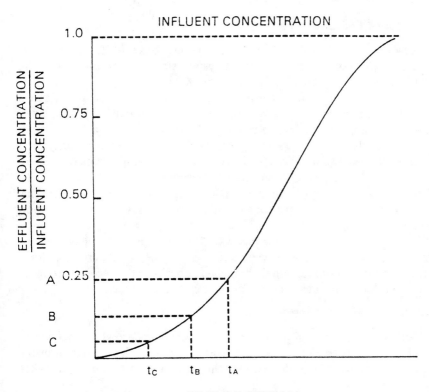

Figure 4. Breakthrough curve and alternative treatment objectives for fixed bed operations.

DESIGN PARAMETERS FOR ADSORBERS

The design of a granular activated carbon adsorption system depends on influent characteristics and treatment objectives (the effluent characteristics). A treatment objective establishes the performance needs for the system; the characteristics of the influent stream affect the choice of system configuration and equipment size necessary to achieve the required performance. Overall economics of the system are heavily dependent on trade-offs between the size of the adsorbers and the size and operating requirements of the carbon reactivation subsystem. System economics also depend on proper design of carbon storage vessels and equipment required for carbon transfer between the adsorption and reactivation subsystems.

Adsorber Characteristics

Adsorbers are cylindrical vessels or rectangular boxes of steel or concrete. The type of vessel used depends on system size, configuration, influent characteristics, and individual designer preference.

The three fundamental adsorber types are open-top steel, gravity flow closed steel, pressure flow, and open-top concrete, gravity flow. Industrial wastewater applications in the range of 3,785–37,850 m³/day (1–10 mgd) would usually employ pressure steel vessels. Large municipal systems greater than 37,850 m³/day (10 mgd) might use open-top concrete adsorbers. Basic characteristics of adsorbers are summarized in Table 3.[10] Other features that differentiate individual adsorbers are the details of internal hardware, such as liquid distributors and collectors and carbon bed support methods.[8,9]

The sizing of an adsorber is based on flow rate, hydraulic loading, and contact time. These variables yield adsorber volume, depth, cross-sectional area, and number of individual parallel adsorber vessels. Once these quantities are known, the adsorber layout plan can be developed.

Adsorber Layout Plans

Varied adsorber layout plans are possible, but standard good engineering practice suggests that certain fundamental approaches be used in all installations. The layout plan depends on the type of adsorber and size of the plant.

Circular Adsorbers

Circular adsorbers include both pressure and gravity flow steel vessels. In at least one instance, circular concrete adsorbers also have been used in a granular activated carbon plant.[11] Use of adsorbers in parallel or in series affects the equipment layout.

A first consideration in developing a layout plan is whether or not to use multiple adsorbers in series. The number of parallel adsorbers or parallel sets of series adsorbers required for a given design capacity must be determined. For the fixed bed case, at least two adsorbers in series is likely. These series pairs would constitute an adsorber module which can be repeated as required to achieve total plant design capacity with parallel modules.

A single adsorber is an unusual situation except for a pulsed bed unit. Since only a portion of exhausted carbon is discharged while the pulsed bed unit continues to operate, a second bed in series, to ensure meeting the effluent requirement while maximizing carbon utilization, is not required. Pulsed bed units can also operate in parallel for large flows.

Selection of parallel circular adsorbers. Selection of the number of adsorbers for any given design depends on vessel size and permissible hydraulic loading. Commercially available steel adsorbers usually have

Table 3. Basic Characteristics of Adsorbers

Adsorber Type	Material	Volume[a] Range (ft³)	Diameter[b]	Remarks
Steel, gravity flow	Lined carbon steel	6,000–20,000	20–30 ft	Field fabricated by welding or bolting preformed steel plate. Mounted on concrete slab foundation.
Steel, pressure flow	Lined carbon steel; stainless steel	2,000–50,000	Up to 12 ft	Shop fabricated; over-the-road transportation constraints limit size.
Concrete, gravity flow	Standard reinforced concrete	1,000–200,000	Usually rectangular	Field constructed. Designs vary. A 2:1 length-to-width ratio is common.

Source: Culp.[10]
[a]To convert ft³ to m³, multiply by 0.0283.
[b]To convert ft to m, multiply by 0.3048.

standard diameters. Shop-fabricated steel vessels are limited to 12 ft in diameter and 60 ft in length by over-the-road transportation constraints. Standard diameters, according to the American Society of Mechanical Engineers (ASME) code for unfired pressure vessels, are given in Table 4 for shop-fabricated and field-fabricated steel vessels.[12]

Rectangular Adsorbers

Rectangular concrete adsorbers are commonly used for municipal drinking water or wastewater facilities that use GAC treatment. Multiple vessels are generally built into a large concrete structure with individual vessels sharing a common wall. Sometimes the overall structure is part of a building foundation. Design details for structural support and influent, effluent, carbon slurry, and backwash inlets and outlets vary from plant to plant. Three major design features subject to variations are the length-to-width ratio of a vessel, the design of influent and effluent channels, and bed depth.

Table 4. Standard Diameters for Circular Steel Vessels

Shop Fabricated[a]	Field Fabricated[a]	
7'11"	15'6"	
9'6"	21'6"	
10'0"	29'9"	unlikely
11'0"	38'8"	for
12'0"	55'0"	adsorbers

Source: Newton, Von Tress, and Bridges.[12]
[a]To convert ft to m, multiply by 0.3048.
  To convert in. to cm, multiply by 2.54.

Table 5. Guide to Selection of GAC System Configuration

Adsorption System Configuration	Applicability
Single adsorber	Low carbon exhaustion rate (usage rate) occurs.
Fixed beds in series	Unusual temperature or pressure conditions must be used in the column.
	High carbon exhaustion rate occurs.
	Gradually sloping breakthrough curve is demonstrated.
	High effluent quality must be assured at all times.
Fixed beds in parallel	High system pressure drop is minimized.
	Large total flow rate is required.
Moving beds, upflow	High carbon exhaustion rate occurs.
	High effluent quality is required. Either some carbon fines can be tolerated in the effluent, or carbon adsorption is followed by filtration.
Expanded beds, upflow	High suspended solids are present in the influent, and some suspended solids can be tolerated in the effluent.
Combined series and parallel	Combined features of respective configurations are required.

A typical flow cross-section for rectangular concrete adsorbers is a 2:1 length-to-width ratio. This is consistent with designs observed in the field and in drawings of actual facilities.

The structural cross-section depends to some extent on the influent and effluent channel design. Structurally, concrete adsorbers can be relatively simple or complex, which obviously influences the cost of a system.

Adsorber Configuration

The two fundamental types of granular activated carbon adsorbers are classified as fixed bed and moving bed. In a fixed bed adsorber, the carbon contained in the adsorber remains stationary and operates in either a downflow or an upflow mode. Moving bed adsorbers, on the other hand, always operate in an upflow mode. An expanded moving bed adsorber operates at a flow rate high enough to cause the carbon bed to expand slightly. A pulsed bed adsorber, the second type of moving bed, is periodically "pulsed" to discharge a portion of the exhausted carbon from the bottom of the bed while a slug of fresh carbon is added at the top. The liquid and carbon move in true countercurrent flow to each other.

Single or multiple adsorbers can be used depending on the application requirements. Multiple adsorbers can be arranged individually in series or in parallel, or as combinations of series units in parallel. Table 5[7,13,14] lists the basic differences in selection criteria for the various adsorber configurations.

Fixed Beds in Series

In this configuration, the flow is downward through the carbon bed for each unit in the series. When carbon is removed for reactivation, the first adsorber in the series is shut down with the next adsorber in line becoming the lead unit. This system can be built with an extra adsorber on standby to become the first adsorber when the lead adsorber is taken out of service. Capital cost considerations usually limit the number of adsorbers in series to four or fewer.[4] Approaching countercurrent flow provides highly efficient usage of the carbon by maximizing the exhaustion of the carbon in the lead adsorber before it is removed for reactivation. Backwashing can be used on the lead adsorber to remove limited amounts of suspended solids that accumulate in the carbon bed.

Fixed Beds in Parallel

In a parallel configuration, each carbon bed receives the same quantity and quality of influent stream. Startup of the individual units is staggered so that exhaustion of the carbon occurs sequentially. This allows removal of all the carbon from each adsorber, one at a time, for reactivation. For systems operating at full design capacity, a spare adsorber can be provided to be brought on line when another adsorber is taken out of service.

In this arrangement, the level of carbon exhaustion when an adsorber is discharged is not as high as that of fixed beds in series because no adsorber is run completely to breakthrough. The carbon utilization in a given adsorber can, however, be increased by blending effluent from all the adsorbers. One or more adsorbers can be run slightly beyond the breakthrough point, while others are not allowed to reach such a high level of carbon exhaustion. This method of operation may be most appropriate for large plants while maximizing carbon utilization.

Expanded Beds

Expanded beds permit removal of suspended solids by periodic bed expansion. In general, expanded beds can tolerate relatively high suspended solids while using finer carbon particle sizes without excessive head loss. Expanded beds cannot be used where downstream contamination by suspended solids or carbon fines passing through the bed would be a problem.

Pulsed Beds

A pulsed bed operates in an upflow mode and the water and carbon flow is countercurrent. Pulsed bed adsorbers permit intermittent or continuous removal of spent carbon from the bottom of the bed while fresh carbon is added at the top without system shutdown. The chief advantage of this system is better carbon utilization because only thoroughly exhausted carbon will be reactivated. In contrast with fixed beds, a pulsed bed cannot be completely exhausted; this prevents contaminant breakthrough into the effluent causing the effluent criterion to be exceeded.

Another characteristic of the pulsed bed system is that a steady-state constant effluent concentration (assuming a constant influent concentration) is achieved. In fixed beds, the effluent concentration gradually increases with time.

Downflow Versus Upflow Operation

Downflow operation is appropriate when the carbon bed is to be used as a suspended solids filter as well as an adsorber. The suspended solids are periodically removed by backwashing in a manner identical to that used with sand media filters. When the carbon adsorber is preceded by sand filtration, downflow operation can sometimes be used without backwashing or with reduced backwashing.

Upflow adsorbers are appropriate for influents with either high or low suspended solids concentrations. For high suspended solids concentrations, upflow adsorbers might be preferred, because solids accumulation and corresponding head losses would be excessive in downflow adsorbers. In such an application, some downstream solids must be tolerated, or the adsorbers must be followed by filtration. For low suspended solids concentrations, upflow adsorbers can be used, because the carbon bed is not needed as a solids filter.

Gravity Flow Versus Pressure Flow

Gravity flow is likely to be used only in downflow beds unless construction on sloping land permits sufficient hydraulic gradient for a gravity-fed upflow.

Pressure flow can be used for either downflow or upflow beds. Pressure flow achieves higher hydraulic loadings (gpm/ft² of adsorber cross-sectional area—m³/min/m²) than would be possible with gravity flow. This higher loading reduces the adsorber cross-sectional area required. Pressure flow also permits operation at higher suspended solids concentrations with no backwashing or with less frequent backwashing than would be possible with gravity flow. This is because there is sufficient head to overcome increasing bed pressure drop (head loss).

Hydraulic Loading

Hydraulic loading is the quantity of liquid flowing through the bed per square foot (gpm/ft²—m³/min/m²). Once the required contact time is elected for a given performance objective, the cross-sectional area of the adsorber is selected to ensure that hydraulic loadings are in a reasonable operating range of 1.35–6.79 L/sec/m² (2–10 gpm/ft²).

The two considerations that determine the permissible hydraulic loading in GAC are superficial bed velocity and head loss. In the normal operating range for hydraulic loading, superficial bed velocity does not appear to critically affect adsorption rates.[15] However, there is evidence that for some materials, changes in superficial velocity do affect performance.[15] This is consistent with the well-established principles of mass transfer from moving fluid streams. Head loss through the carbon bed is a more dominant concern in the selection of hydraulic loading.

Estimation of head loss is discussed in more detail elsewhere.[8,15] In addition to flow rate and influent solids concentration, head loss also depends on carbon particle size and type. Figure 5 shows clean bed head loss curves for several commercially available carbons. The permissible head loss also depends on whether gravity or pressure is used in downflow.

Because of head loss, the acceptable loading for any liquid stream depends on the quantity and characteristics of suspended solids present and their tendency to plug the bed. Downflow beds are more subject to plugging than upflow beds. In an upflow mode there is a tendency for particles to settle away from the upstream end of the bed (since in upflow, upstream is the bottom of the bed), thus preventing surface binding and penetration of solids into the bed. Also, upflow beds are easily expanded so that solids can wash through. In a downflow mode, solids accumulate at the surface and within the bed and bed expansion can be achieved only in a backwash cycle. In some applications, the carbon is used deliberately as a filter as well as an adsorber, in which case the downflow arrangement is preferred. Commonly, carbon adsorption is preceded by sand media filtration to reduce suspended solids concentration and resulting head losses.

Typical ranges for hydraulic loading are 1.35–4.07 L/sec/m^2 (2–6 gpm/ft^2) for downflow beds and 3.40–6.79 L/sec/m^2 (5–10 gpm/ft^2) for upflow beds. It has been recommended that for gravity downflow, hydraulic loading should be limited to less than 2.72 L/sec/m^2 (4 gpm/ft^2), but for pressure downflow, this maximum value can be increased. For upflow expanded beds, 4.07 L/sec/m^2 (6 gpm/ft^2) for 8 × 30 mesh carbon is acceptable. Other recommendations based on system configuration are given in Table 6.

Backwashing

Backwashing is the reverse flow of treated effluent through a carbon bed to dislodge and remove accumulated suspended solids. Frequency of backwashing depends on the rate of head loss buildup, and the rate of increase in effluent suspended solids concentration. The quantity of backwash water depends on the amount and characteristics of the accumulated solids. Backwash frequency and quantity depends, therefore, on the interaction of hydraulic loading and other system variables previously discussed.

The frequency of backwash must be selected to ensure adequate solids removal and prevention of compaction of accumulated solids. A guide to

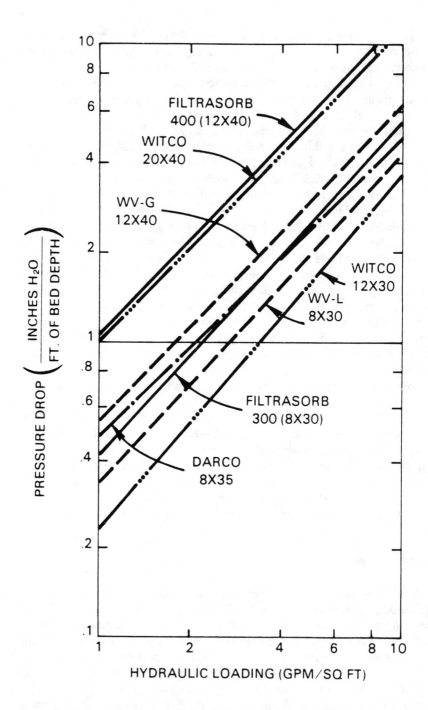

Figure 5. Carbon bed pressure drop vs hydraulic loading.

Table 6. Typical Hydraulic Loadings for Different Adsorber System Configurations

Configuration	Hydraulic Loading (gpm/ft²)ᵃ
Parallel and single column fixed bed	1–4
Series beds	3–7
Pulsed bed systems	5–9

Source: Culp.[10]
ᵃTo convert from gpm/ft² to L/sec/m², multiply by 0.679.

selecting backwash frequency and duration is that no more than 5% of the treated clean water should be required for backwashing.[8]

Supplements to backwashing are air scouring and surface washing. Air scouring is the introduction of compressed air at the bottom of the bed during backwashing to increase turbulence and loosen deposited solids. Its benefits must be weighed against the potential for increased carbon losses due to abrasive attrition. Surface wash is the application of water jets to the upstream surface of the bed to loosen compacted solids. Systems with backwash do not necessarily use these supplementary operations.

REACTIVATION SYSTEMS

Three potential reactivation techniques are chemical, steam, and thermal. Both chemical and steam reactivation are used primarily in industrial systems designed to recover the adsorbate. These techniques have not been used in water and wastewater treatment systems, since they do not adequately reactivate the carbon for uses requiring removal of organics present in low concentrations. Thermal reactivation is the reactivation method commonly used in water and wastewater treatment systems; therefore, it is the only method which will be discussed in detail here.

Thermal reactivation is made up of four stages: drying, desorption, pyrolysis, and gasification. The granular carbon is usually transported to the reactivation equipment in a water slurry and is dewatered to about 50% water by weight in a dewatering feed screw to the reactivation furnace. Drying occurs in the entry zone as the carbon is heated from ambient temperature to 100°C (212°F). Thermal desorption followed by pyrolysis of the organic adsorbate occurs as the temperature increases beyond the drying temperature. Gasification occurs between 649°C (1200°F) and 1038°C (1900°F).

In the desorption step, volatile materials that are not strongly adsorbed to the carbon are driven off as they are heated; then, heavy organics are pyrolized, forming residual char on the carbon. Finally, during the gasification stage, the desorbed vapors and gaseous products of pyrolysis leave the pores of the particles. During this stage the char is burned. Reactivated

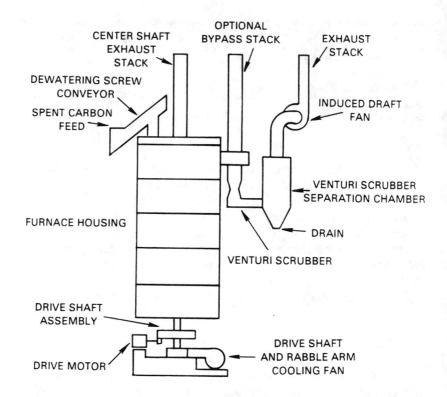

Figure 6.   Typical multiple hearth furnace regeneration system.

carbon generally does not have the same adsorption characteristics as new carbon, due partly to enlargement of the pore sizes during pyrolysis and the adsorption of some inorganic compounds from the water that are not removed in reactivation. After the reactivation is completed, the hot carbon is cooled in a water-filled quench tank.

Four different thermal reactivation systems are rotary kilns; multiple hearth furnaces; fluid bed furnaces; and infrared furnaces. Multiple hearth is the traditional method of granular activated carbon reactivation, although fluidized bed and infrared furnaces are being applied more frequently. Because they are more energy intensive, rotary kilns do not appear to be competitive with the other technologies.

Multiple Hearth Furnaces

Figure 6 shows a drawing of a typical multiple hearth furnace system. Figure 7 shows an internal cross-sectional view of a multiple hearth furnace. This type of furnace consists of a refractory-lined steel shell containing five to eight circular hearths. Burners are generally located on the bottom hearth to provide necessary heat for reactivation and can also be located on higher

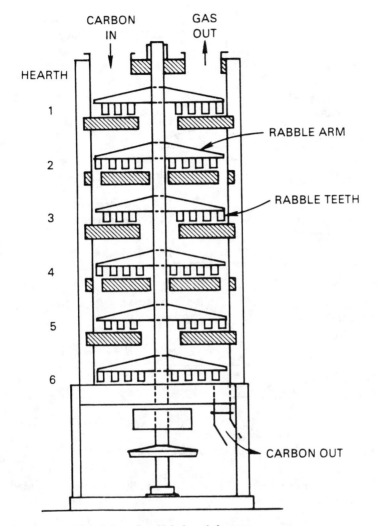

Figure 7. Cross-sectional view of multiple hearth furnace.

hearths to provide improved temperature profiles for increased performance and flexibility of operation.

   Spent carbon enters the top of the furnace through a dewatering screw
conveyor onto the top hearth. A rotating center shaft supporting arms with
rabble blades moves the carbon in a spiral path so that it drops from hearth
to hearth and finally out of the bottom of the furnace into a quench tank
for cooling. Residence time in the furnace is controlled by the rotation
speed of the center shaft. The hollow rabble arms are cooled by ambient air
blown through them. The atmosphere within the furnace is tightly con-

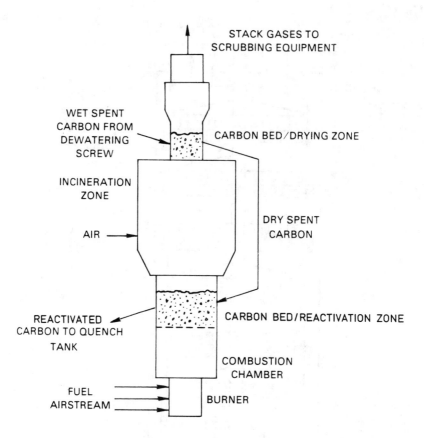

Figure 8.   Fluidized bed regeneration furnace.

trolled to prevent excessive carbon oxidation. Steam is usually added in the lower hearths.

Fluid Bed Furnaces

A fluid bed furnace suspends the carbon particles by an upward flowing gas stream. The velocity of the gas is controlled so that the weight of particles in the bed is just balanced by the upward force of the gas. The use of fluid bed in this application offers the advantages of uniform temperatures within the bed and high heat and mass transfer rates. A schematic of the reactivator is shown in Figure 8. Dewatered carbon is dried in the first zone. Steam must be injected to control the temperature.

Carbon leaves the first zone with a moisture content of about 1%. Dried carbon flows by gravity to the bottom zone; the lower zone is maintained at about 982°C (1800°F). As the carbon temperature increases to the temperature of the zone, adsorbed organic materials undergo the reactivation steps of pyrolysis and gasification. After leaving the furnace, the carbon is cooled in a water-filled quench tank.

Air is injected into the space between the two zones, which serves as a second combustion zone. This burns the organic compounds, and the hydrogen and the carbon monoxide that are released during reactivation in the lower chamber. Part of the heat produced dries the wet carbon in the upper bed, with the remainder recycled to the lower bed reactivation zone. Addition of air at this location reduces or eliminates the need for an after-burner. If an afterburner is not needed, less energy is required than that required by a multi-hearth furnace. If there is no afterburner, the gases leaving the fluid bed regenerator will be cooler than those produced by multi-hearth furnaces.

Infrared Furnaces

An infrared furnace consists of an insulated enclosure through which the carbon is transported on a continuous metal conveyor belt. Heat for reactivation is supplied to the furnace by a number of electric infrared heating elements mounted in the top of the tunnel.

Spent carbon is fed to the reactivator through an airlock that controls the carbon feed rate and minimizes air intake. Carbon is spread evenly ($3/4$ –1 in.; 1.90–2.54 cm) onto the belt and transported slowly through discrete chambers or zones of the furnace. The chambers provide increasingly higher temperatures and accomplish the four reactivation stages discussed previously. Atmosphere and temperature can be tightly controlled with this type of furnace. Carbon discharges into a conventional water-filled quench tank for cooling.

Figure 9 shows a cross-sectional view of an infrared furnace developed by Shirco, Inc., which is the only known company to have commercialized this process to date. The unit is built of standard modules that allow for quick assembly and startup. Modules are made of a steel shell lined with a thermal shock–resistant ceramic fiber blanket insulation and support rollers for the conveyor belt.

Reactivation Design Parameters

Reactivation Rate

The first consideration in the design of a reactivation system is the reactivation rate, which depends on carbon loading and carbon usage or exhaustion rate in the adsorbers. Carbon loading is the quantity of organic substance removed from the water per unit quantity of carbon. Carbon loading is usually expressed in terms of milligram per gram (mg/g) or pounds per pound (lb/lb – kg/kg) of organic materials to carbon. This is not the equilibrium carbon loading determined by isotherm tests; rather, it is the value

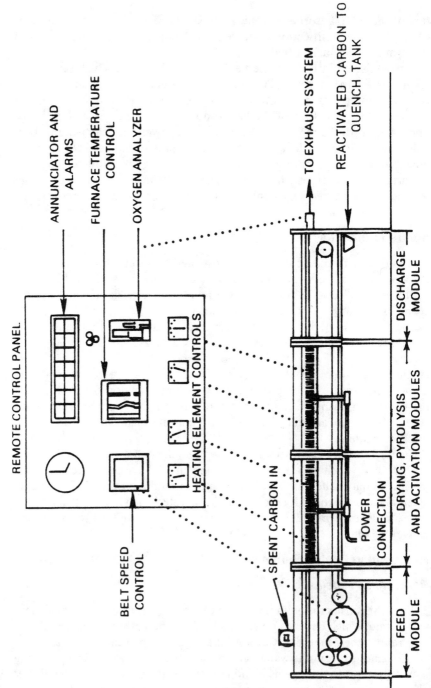

Figure 9. Infrared regeneration furnace.

Table 7. Typical Overall Parameters for Three Reactivation Systems

	Multi-hearth	Fluidized Bed	Infrared
Furnace loading (lb/ft²/day)	40–115	1460–1700	—
(kg/m²/day)	(195.2–561.2)	(7124–8296)	
Residence time (min)	120–240	10–20	25
Natural gas fuel (scf/lb C)	3.0–5.5	1.5–2.2	—
Electricity (kwh/lb C)	—	—	—
Steam (lb/lb C)	1.0–3.0	0.4–0.8	—
Heat loss (%)	15	9	—
Carbon losses (%)	5.7	2–5	5

— = Data not available.

achieved with a given system under given operating conditions. An increase in carbon loading for the same treatment objective lowers reactivation carbon throughput in the furnace due to the longer residence time required to reactivate the carbon. Carbon usage is sometimes expressed as carbon dosage or pounds of carbon per million gallons of water treated. It is inversely related to carbon loading.

Carbon usage rate, or exhaustion rate, is the mass of carbon exhausted per unit time. It is usually expressed as pounds per day (lb/day − kg/day) and is inversely related to reactivation frequency, which measures how often a given bed of carbon must be reactivated. It is the minimum rate at which carbon would be reactivated in a continuous system with no intermediate storage. In practice, carbon storage is placed between the adsorbers and reactivator so that the furnace sizing is based on higher flow rates of carbon than this minimum value. Operation at the higher rate compensates for furnace downtime while still meeting the carbon usage rate of the adsorbers.

Furnace Size

Furnace size is based on the rate at which carbon is charged per unit of furnace dimension and varies with the type of reactivation furnace chosen. For multiple hearth furnaces, the value is usually given in pounds of carbon per square foot of hearth area per day (lb/ft²/day − kg/m²/day). For multi-hearth furnace design, the values have been reported in a range of 145–561 kg/m²/day (40–115 lb/ft²/day). Values for fluidized bed furnace loadings range from 7125 to 8296 kg/m²/day (1460–1700 lb/ft²/day). These numbers are based on the cross-sectional area perpendicular to the flow of combustion air. In an infrared furnace, carbon flow depends on the width of the conveyor belt and residence time. Table 7[16,17] lists typical overall parameters for the three furnace types.

Temperature and Atmosphere Control

Temperature and composition of the vapor space in the furnace are important operating parameters. These two variables influence the amount of carbon that is lost during reactivation. To minimize carbon loss, it is important to maintain the proper amount of oxygen in the furnace. The oxygen content must be high enough to pyrolyze the char but not so high as to burn the carbon. At a fixed operating air moisture, the temperature must be controlled for the same reasons. A precise temperature profile exists for each type of GAC. Experience with industrial and municipal wastewater treatment indicates that losses can be held to 8–10% per cycle and the adsorptive activity of the reactivated carbon is 90% of the virgin activity by proper adjustment of furnace variables.[15]

Offgas Control

The offgases resulting from the reactivation process contain potential air pollutants. The most common approach to offgas pollution control is to use an afterburner followed by a wet scrubber. The afterburner oxidizes organic compounds, and the scrubber removes particulate matter and any soluble chemical species from the gas stream.

Many different scrubber designs are used for cleaning reactivated offgas. Water is the usual scrubbing medium. Infrared furnaces require the smallest amount of scrubber water circulation of all the furnaces, due to the relatively low offgas flow rate. Multi-hearth furnaces probably use the most scrubber water because of the higher offgas temperatures.

A supplementary dust collector is used only where it is necessary to collect particulate matter not normally removed by scrubbers.

Offsite Custom Reactivation

One alternative to reactivating GAC at the plant site is custom offsite reactivation. Spent carbon is shipped to another location, where it is batch-reactivated to minimize intermixing with other carbons. Reactivated carbon is then shipped back to the treatment plant for use. Currently, offsite reactivation is a contracted service for which the user pays a fee based on the amount of carbon involved.

CARBON STORAGE AND TRANSFER

Carbon storage and transfer includes those parts of the granular activated carbon plant associated with fresh carbon receipt, unloading, and transfer to the adsorbers; spent carbon removal from the adsorbers and transport to the reactivation subsystem; and reactivated carbon transport back to the adsorbers. Commerical granular activated carbon is shipped in one of several ways: bags or drums by truck or rail car; bulk by truck; or bulk by rail car. The choice of shipment method and design of unloading facilities depend on carbon usage requirements and local shipment economics. In general, however, individual container shipment (bags or drums) would be used for only the smallest systems.

Carbon vendors use special bulk trucks which discharge a slurry at the plant site. Piping and transfer hoses are part of the truck. When carbon is shipped in bulk by truck or rail car, direct slurry unloading can be used. This can be done through hoses and special connections on the shipment vehicles or from a dry dump station with slurry makeup directly into a storage vessel. The method used will depend on the quantity and frequency of the shipments of fresh carbon to a specific plant. The choice between open or sheltered storage depends on the need for protection from the weather.

The methods for transferring fresh carbon into the adsorbers determine the kinds of equipment required. A common method is to prepare a carbon slurry in a makeup tank and transfer the slurry to the adsorbers when required. The quality and frequency required for each carbon slurry batch depends on the makeup carbon needs of the specific installation.

Spent carbon is discharged from the adsorbers as a slurry and usually travels to a spent carbon storage tank rather than directly to the reactivation subsystem. After reactivation, the carbon returns as a slurry to a reactivated carbon storage tank prior to returning to the adsorbers.

Hydraulic transfer of the carbon slurry is accomplished by one of three devices: eductor, slurry pump, or blowcases. Eductors are standard vendor-supplied items. A centrifugal pump supplies water to the eductor venturi and the slurry is drawn into the venturi by pressure differential, mixed with water, and moved down the pipeline.

Slurry pumps are centrifugal pumps with solids-handling open impellers. Details of these pumps vary according to manufacturer. Blowcases are small pressure vessels into which a carbon slurry may be transferred by some other method. The blowcase is pressurized with compressed air, which forces the carbon slurry into the outlet line leading to the transfer destination.

Spent carbon is typically transported in a concentration of up to 0.1 kg of carbon per liter of water (3 lbs/gal).[11] This concentration permits the use of slurry pumps, avoids eductor plugging, and minimizes pressure drop, pipe size, and attrition losses. Attrition losses occur when carbon abrasion between particles generates fines, which are removed from the system during backwash or regeneration.

CONSTRUCTION MATERIALS

Materials of construction for granular activated carbon plants vary. Steel, concrete, or fiberglass are used for vessels, and steel, plastic, or fiberglass for piping. Individual influent stream characteristics and type of carbon used determine construction material requirements.

Dry carbon and carbon in water slurries are not corrosive. Damp carbon, however, is extremely corrosive.[8] To overcome this corrosivity, all metal that will be in contact with damp carbon has to be corrosion-resistant.

Steel adsorbers are usually made of carbon steel coated with rubber, epoxy, or other materials. It has also been suggested that steel tanks can be protected by a cathodic protection system.[8] Solid stainless-clad carbon steel may be useful, especially for industrial systems. Fiberglass vessels are another option for overcoming the corrosion problem. Corrosion problems do not occur with concrete vessels except for metal parts such as nozzles, screens, piping, and weirs.

Piping can be either carbon steel or stainless steel, depending on its location and purpose in the system. As with vessels, various linings can be used for piping. Valves and instruments such as flow meters must be of materials compatible with the potential for corrosion in their respective parts of the system. Using carbon steel with periodic replacement might be less costly than stainless steel construction.[12] Plastic or fiberglass piping offers another alternative where corrosion problems are anticipated.

In reactivation systems, both carbon steel and stainless steel are used for various equipment components. Furnaces are usually carbon steel with refractory liners. Certain furnace components in contact with damp carbon or exposed to high temperatures are made of stainless steel. Dewatering feed screws and quench tanks are examples of such items.

REFERENCES

1. Cheremisinoff, P. N., and F. Ellerbusch, Eds. *Carbon Adsorption Handbook* (Ann Arbor, MI: Ann Arbor Science Publishers, Inc., 1978).
2. Culp, R. L., G. Wesner, and G. L. Culp. *Handbook of Advanced Wastewater Treatment* (New York: Van Nostrand Reinhold Co., 1978).
3. Weber, W. J. "Adsorption, Introduction and General Rules," in *Calgon Adsorption Handbook* (Pittsburgh, PA: Calgon Corporation, undated).
4. Hager, D. G. "Source Treatment of Industrial Wastewater with Granular Activated Carbon," unpublished from manuscript entitled "Economics of Adsorption in Wastewater Treatment," paper presented at the Applications of Adsorption to Wastewater Treatment Conference, Vanderbilt University, Nashville, TN, February 16–19, 1981.

5. Weber, W. J., and J. C. Crittenden. "MADAM I-A Numeric Methods for Design of Adsorption Systems," *J. Water Pollut. Control Fed.* 47(5):924 (1975).
6. Perry, R. H., C. H. Chilton, and S. D. Kirkpatric. *Chemical Engineer's Handbook*, 4th ed. (New York: McGraw-Hill Book Co., 1973), pp. 16–18.
7. Rizzo, R. L., and A. R. Shepherd. "Treatment of Industrial Wastewater with Activated Carbon," *Chem. Eng.* (January 1977).
8. "Process Design Manual for Carbon Adsorption," U.S. EPA Technology Transfer (October 1973).
9. Carnes, W. A. "Alternative GAC Adsorber Design," unpublished report, Zurn Industries, Inc., Dallas, TX.
10. Culp, R. L. "GAC Water Treatment Systems," *Public Works* (February, 1980), pp. 83–87.
11. Zurn Industries, Inc. "Upflow-Downflow Carbon Adsorption," paper presented at the 68th meeting of the American Institute of Chemical Engineers.
12. Newton, P., W. R. Von Tress, and J. S. Bridges. "Liquid Storage in the CPI," *Chem. Eng.* (Deskbook Issue, April 1978), pp. 10–11.
13. Zanitach, R. H. "Application and Engineering Design Considerations," in *Calgon Adsorption Handbook*, (Pittsburgh, PA: Calgon Corporation, undated).
14. Fornwalt, H. J., and R. A. Hutchins. "Purifying Liquids with Activated Carbon," *Chem. Eng.* (April 1966).
15. Hutchins, R. A. "Liquid-Phase Adsorption: Maximizing Performance," *Chem. Eng.* (February 1980).
16. Guarino, C. F., J. V. Radzini, R. R. Cairo, D. D. Blair, M. M. Pence, B. S. Aptowicz, and N. Weintraub. "Design and Economic Considerations of GAC Systems," *Water Sew. Works* (September 1980), pp. 58–57.
17. Smith, S. B. "Activated Carbon III — Alternatives and Relative Costs of Regeneration Processes," *Public Water Supply Eng.* (Deskbook Issue, October 1977), p. 104.

CHAPTER 3

EPA's GAC Field-Scale Studies

INTRODUCTION

Based on recommendations by the National Drinking Water Advisory Council, EPA's Drinking Water Research Division initiated a series of field-scale projects designed to study the cost and performance of granular activated carbon.

Ten utilities participated in cooperative GAC research efforts. Listed below are the locations of these utilities and some of the objectives established for each.

- Cincinnati, Ohio
  1. Determine if the use of granular activated carbon utilizing either deep bed contactors or conventional depth gravity filters with onsite GAC reactivation is feasible for removing specific trace organics from Ohio River water while treating it for human consumption. Bench-scale pilot columns were compared to full-scale deep bed, postfiltration activated carbon contactors (adsorbers), and conventional depth sand replacement carbon beds (filter adsorbers).
  2. Evaluate onsite fluidized bed GAC reactivation for cost of operation and reactivation performance.
  3. Develop plant design and operating parameters for full-scale plant conversion to GAC treatment.

- Manchester, New Hampshire
  1. Evaluate the cost and performance of a GAC reactivation system utilizing fluid bed principles.
  2. Evaluate the cost and performance of a semiautomatic system for hydraulically transporting GAC to and from its point of reactivation.

3. Determine the adsorptive effectiveness of both virgin and reactivated GAC for removing TOC from Manchester drinking water.

• Jefferson Parish, Louisiana (JP-1)*
1. Provide data on the efficiency of GAC bed adsorption for a wide range of organic chemicals. This included data on the frequency at which hydrocarbons enter the process stream from the raw water source. Also, removal of organic species by various treatment units or processes was investigated, such as lime softening, polyelectrolyte treatment, addition of potassium permanganate, flocculation, filtration, postfiltration adsorption, and combined filtration adsorption (filter adsorber).
2. Investigate certain nonspecific parameters such as TOC for use as surrogate monitoring parameters for GAC applications.
3. Compare the relative efficiency of postfiltration adsorption vs. combined filtration adsorption.

• Evansville, Indiana
1. Develop a water treatment process using chlorine dioxide as a disinfectant to evaluate the resultant production of any organic compounds and their removal by GAC. Also, determine the effectiveness of virgin and subsequently reactivated GAC for removal of any organics, whether present in the source water or formed during treatment.

• Miami, Florida
1. Conduct bench-scale study on the effectiveness of two adsorbent resins and one type of GAC in removing various organic compounds from raw and treated drinking water.

• Ohio River Sanitation Commission (ORSANCO): Huntington, West Virginia and Beaver Falls, Pennsylvania
1. Investigate and evaluate full-scale GAC treatment for control of organic compounds.

• Passaic Valley, Little Falls, New Jersey
1. Evaluate the efficiency of three types of GAC before and after infrared reactivation.

• Philadelphia, Pennsylvania
1. Investigate the effectiveness of GAC in replacing filter sand for removal of a wide range of organic chemicals from a contaminated source water.
2. Evaluate the use of ozone in combination with GAC to increase the efficiency of organic contaminant removal.

• Wausau, Wisconsin
1. Evaluate the effectiveness of GAC for removing a large number of volatile organic compounds from a contaminated groundwater.
2. Develop protocols for the use of mathematical models to provide useful information for full-scale design of GAC contactors.

---

*Two GAC research projects were started at Jefferson Parish, Louisiana. Only the first project (JP-1) will be discussed.

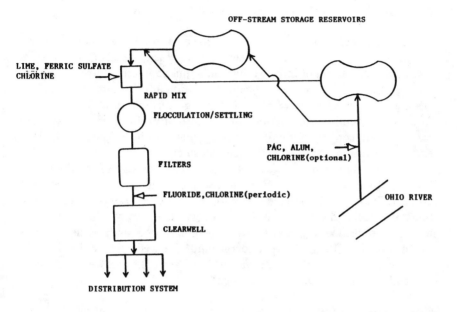

Figure 1.   Schematic of Cincinnati, Ohio waterworks.

• Thorton, Colorado
  1. Evaluate the effectiveness of GAC after coagulation/settling (filter adsorber) and after sand filtration (adsorber).

With the exception of Wausau, each full-scale facility involved in the GAC research effort utilized essentially the same treatment scheme: predisinfection, coagulation/flocculation, settling, filtration, and postdisinfection. Minor variations of this treatment scheme were practiced by a few utilities. A general description of each full-scale and pilot plant operation is presented below as background information in relationship to the performance data that will be presented later.

CINCINNATI, OHIO

The primary source water for the Cincinnati Water Works is the Ohio River (Figure 1).[1] Raw water is pumped into large presettling basins having a combined capacity of 1.4 million m³ (372 million gal) and a retention time of two to three days. Alum at an average dose of 17 mg/L was added to the raw water to aid settling. Occasionally, powdered activated carbon was added to the raw water as a sunlight screen to deter algae growth in the presettling basins (taste and odor control) and to lessen the effect of a chemical spill.

From the presettling basins, water flowed through two parallel water turbine–powered generators that serve as a velocity breaker. The turbine discharge flow was split into two hydraulic jumps, thereby providing rapid mixing of chemicals. Average chemical addition at this point included lime (17 mg/L), ferric sulfate (8.6 mg/L for high turbidity and 3.4 mg/L for filter conditioning during periods of low turbidity), fluoride (to produce a plant effluent concentration of 0.9 mg/L), and chlorine (typical plant effluent concentration was 1.8 mg/L free chlorine).

The hydraulic jumps directed water to two parallel basins for flocculation followed by clarification. Each basin had a capacity of 49,215 m³ (13 million gal). Two flumes directed water to 47 rapid sand gravity filters (gradations of gravel and sand), each having an effective area of 130 m² (1400 ft²) with a hydraulic loading of 6.0 m/hr (2.5 gpm/ft² or 5 mgd). Finished water was directed to either of two underground clearwells before distribution. Total design flow of the plant was 10.3 m³/sec (235 mgd) with a typical demand of about 5.7 m³/sec (130 mgd).

Modification of Existing Sand Filters

During the GAC study, four of the water treatment plant's 47 sand filters were converted to GAC filters. One other sand filter was used for storage of makeup GAC. Because the "Ten State Standards" require a minimum of 30.5 cm (12 in.) of filter sand for turbidity removal, only 45.7 cm (18 in.) of one filter's 76.2 cm (30 in) of sand was removed and replaced with Westvaco 12 × 40 WVG® GAC.[2]

With the concurrence of the Ohio EPA, all of the sand (76.2 cm, 30 in.) in three other filters was replaced with GAC. Westvaco 20 × 50 WVW® GAC was used in one of these filters to coincide with the effective size (0.45 to 0.55 mm) of the sand. This filter and the one with 45.7 cm (18 in.) of GAC were used as post–sand filtration adsorbers. The other two filters contained 76.2 cm (30 in.) of Westvaco 12 × 40 WVG GAC and were used as filter adsorbers, receiving water before sand filtration (after coagulation/settling). Average empty bed contact times for all of the modified filters were around 7.5 minutes.

GAC Contactors

The performance of the previously described conventional depth (76.2 cm, 30 in.) gravity filters was compared to deep bed contactors (4.6 m, 15 ft). Four contactor tanks were utilized for this comparison. Each contactor was of cylindrical shell construction of 9.55-mm (0.375 in.)–thick carbon steel designed to the following specifications:

diameter	3.4 m (11 ft)
vertical sidewall height	6.7 m (22 ft)
GAC depth	4.6 m (15 ft)
design capacity	0.04 m^3/sec (1.0 mgd)
hydraulic loading	17.8 m/hr (7.4 gpm/ft^2)
contact time	15.3 min
design pressure	517.1 kPa (75 psig)
test pressure	620.5 kPa (90 psig)
backwash hydraulic loading	24 m/hr (10 gpm/ft)

The GAC used in each contactor was 12 × 40 WVG.

GAC Pilot Columns

Bench-scale glass columns 7.6 cm (3 in.) in diameter were used to simulate the sand replacement filter beds. Two columns 1.4 m (54 in.) in height and containing 76.2 cm (30 in.) of 12 × 40 WVG were utilized. In addition, pilot columns were used for performance comparison to the deep bed contactors. One set of four columns in series contained ICI Hydrodarco® 10 × 30 GAC. Each of these columns was 10.2 cm (4 in.) in diameter and 1.8 m (6 ft) in height. The first column in the series had 0.9 m (3 ft) of GAC and each of three successive columns had 1.2 m (4 ft) of GAC for a total GAC depth of 4.6 m (15 ft).

Another series of columns was comprised of four 1.5-m (5-ft) sections 10.2 cm (4 in.) in diameter. These columns contained WVG 12 × 40 GAC. The first column consisted of 0.9 m (3 ft) of GAC, while each of the three succeeding columns contained 1.2 m (4 ft) of GAC, giving an aggregate GAC depth of 4.6 m (15 ft).

MANCHESTER, NEW HAMPSHIRE

Lake Massabesic serves as the principal water source for the city of Manchester (Figure 2).[3] This lake supply is of natural origin, having a watershed of 114 km^2 (44 mi^2). Water treatment is accomplished in a 1.8-m^3/sec (40-mgd)–capacity facility.

The treatment process consisted of five sequential unit processes beginning with flash mixing, followed by flocculation, sedimentation, sand filtration, and GAC adsorption. Water was delivered to the flash mix basin by a low lift intake station located on the shore of Lake Massabesic. In the flash mix basin, alum and sodium aluminate were added for coagulation, pH adjustment, and alkalinity control at dosage levels averaging about 12 mg/L and 8 mg/L, respectively. Water flowed through a common distribution channel to four covered flocculation tanks. Flocculation detention time was about 1.3 hr at an average day finished water flow of 9.6 m^3/sec (13 mgd).

Water from the flocculation tanks passed through a row of baffles to four covered sedimentation basins. Sedimentation detention time averaged

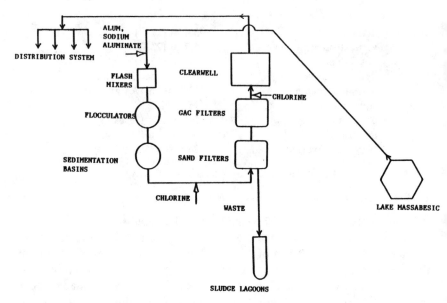

Figure 2.   Site plan—Manchester, New Hampshire water treatment plant.

about 5 hr at the average flow. Following sedimentation, water flowed over a series of serpentine-type weirs into a common effluent channel, where it was directed into the filter room. Filtration was accomplished by four 27.9-cm (11-in)–deep rapid sand filters (effective size 0.60 to 0.65 mm) followed by adsorption on four GAC filters containing 1.2 m (48 in.) deep beds of Westvaco WVW 8 × 30 carbon. Both the sand and GAC filters were 4.9 m (16 ft) in width by 33.5 m (110 ft) in length. All of the filters contained compartmented filter cells one foot in width. The sand filters were equipped with automatic backwash (ABW) filter carriages that traveled the length of the filter individually backwashing the cells without taking the filter out of service. Average hydraulic loading was 3.4 m/hr (1.4 gpm/sq ft) and the EBCT for each GAC filter was about 22 min.

Disinfection was accomplished by pre- and postchlorination. Prechlorine was added just before sand filtration at an average dose of 1 mg/L. Post-chlorine was added at the clearwell in the range of 2.0 to 3.0 mg/L to produce an average distribution free chlorine residual of 0.5 mg/L.

Full-Scale Evaluation

In order to evaluate GAC during full-scale application, one of the four GAC filters normally used for taste and odor control was utilized as a test filter. At the beginning of the study, one-half of the filter contained virgin WVW 8 × 30 mesh GAC and the other half once-reactivated GAC (5-year-old service carbon reactivated at the beginning of the study). A complete

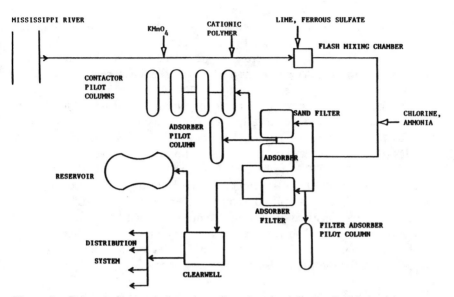

Figure 3.   Full-scale filter and pilot column flow chart for Jefferson Parish, Louisiana.

separation of the GAC was possible, because the GAC filters contained 110 cells spaced at 30-cm (12-in.) centers.

JEFFERSON PARISH, LOUISIANA

The conventional treatment plant at Jefferson Parish pumped raw source water from the Mississippi River (Figure 3).[4] Total design flow of this plant was 3.1 m³/sec (70 mgd) utilizing four separate smaller treatment plants with a demand of about 2.4 m³/sec (55 mgd). Only one of the smaller plants within the treatment complex was used for this research effort at a design flow of 0.4 m³/sec (10 mgd).

Potassium permanganate (0.5–1.0 mg/L) was added for taste and odor control as the first treatment step. A cationic polyelectrolyte, diallyldimethyl diammonium chloride (0.5–8.0 mg/L) was then added as the primary coagulant. Lime (7–10 mg/L) was continuously fed for pH adjustment to 8.0–8.3. After the precipitators, chlorine and ammonia were added in a ratio of 3:1 for chloramine disinfection before sand filtration. Monochloramine concentrations prior to the adsorbers and filter adsorbers ranged from 1.4–1.7 mg/L.

Although the above presents the general treatment scheme, some deviation did occur. On occasion, ferric chloride was added as a coagulant aid in an attempt to reduce the high cost of the polyelectrolyte. The addition of lime was discontinued at one point and sodium hexametaphosphate added for corrosion control in the sand filter system.

Two sand filters were converted to granular activated carbon beds for evaluation. One was designated as a filter adsorber because it served a dual purpose of removing filterable turbidity and adsorbing dissolved organic compounds. This filter received clarified chloraminated water. Media for the filter adsorber consisted of a 15.2-cm (6-in.) layer of sand covered by 61 cm (24 in.) of GAC in two modes of operation (I and IIA). During the other modes (IIB and III), the sand layer was removed so that the filter adsorber contained 76.2 cm (30 in.) of GAC. For modes I, IIA, and IIB, the GAC was Westvaco WVG 12 × 40 and for mode III, Calgon Filtrasorb® 400.

The second filter was utilized as an adsorber for post–sand filter adsorption. This filter consisted of 76.2 cm (30 in.) of the same type of GAC used in the filter adsorber. The average hydraulic loading for these filters was 2.2 m/hr (0.9 gpm/ft²) with corresponding EBCTs of 17.5 and 20.4 min, respectively, for the filter adsorber and adsorber.

Pilot Columns

Two GAC pilot columns were operated under the same conditions and in parallel with their full-scale counterparts, filter adsorber and adsorber. These pilot columns were given the designations of adsorber simulator and filter adsorber simulator. Each column received influent water from the same sources and at the same relative hydraulic loadings as the full-scale units.

The pilot column simulators were constructed of glass 7.6 cm (3 in.) in diameter and 1.5 m (5 ft) in height, and contained 76.2 cm (30 in.) of GAC on top of 15.2 cm (6 in.) of gravel support. Also, four other GAC pilot columns were operated in series at a flow rate of 0.76 L/min (0.2 gpm) to simulate the effects of 10, 20, 30, and 40 min of EBCT on postfiltration contactors. These glass columns were 10.2 cm (4 in.) in diameter and 1.8 m (6 ft) in height and contained 91.4 cm (36 in.) of GAC on top of 15.2 cm (6 in.) gravel support. The influent water for these pilot postfiltration contactors was the same as that for the adsorber filter and adsorber simulation column.

EVANSVILLE, INDIANA

The Evansville Waterworks Filtration Plant uses the Ohio River as source water with intakes about 1.6 km (one mile) upstream from the city of Evansville (Figure 4).[5] This plant consists of two separate treatment systems (North and South Plants) with each having a hydraulic capacity of approximately 1.3 m³/sec (30 mgd). The South Plant was used as a control while evaluating the performance of an onsite pilot plant.

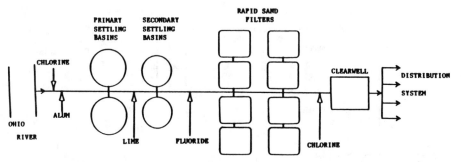

Figure 4.   Flow diagram of Evansville, Indiana full-scale South Plant.

The South Plant consisted of two primary settling basins, two secondary settling basins, and eight rapid sand filters. Each primary settling basin had a capacity of approximately 6813 m³ (1.8 million gal). The secondary settling basin capacities were 2744 m³ (0.725 million gal).

Chlorine and alum were added before primary settling with average concentrations of 6 and 28 mg/L, respectively. A free chlorine residual of 1.5–2.0 mg/L was maintained after sand filtration, with any makeup chlorine being added before the water passed into a common clearwell. Approximately 12 mg/L of lime was added after primary settling for pH adjustment to 8.0 and about 1.5 mg/L of fluoride was added immediately before the rapid mixed-media filters.

Pilot Plant

A 0.35-m³/min (100-gpm) pilot plant with a detention time of approximately 37 min was used for extensive evaluations (Figure 5). Disinfection was accomplished by onsite generation of chlorine dioxide. Average alum and polymer dosages of 12 and 0.8 mg/L, respectively, were added to the raw water for turbidity control. For pH control to about 8.0, an average lime dosage of approximately 6.0 mg/L was added prior to filtration.

The pilot plant consisted of a rapid mix chamber, flocculator, tube settlers, and mixed-media filter. Two post–pilot plant GAC contactors were used that had a column height of 2.7 m (9.0 ft), an inside diameter of 0.97 m (38 in.), and an average GAC bed depth of 2.0 m (6.5 ft). An EBCT of 9.6 min was established with a hydraulic loading of 12.2 m/hr (5.1 gpm/ft²).

The GAC used in this study was ICI Hydrodarco 10 × 30.

MIAMI, FLORIDA

The John E. Preston Water Treatment Plant, located in Hialeah, Florida, is one of three water plants serving the greater Miami area (Figure 6).[6] This plant was operated by the Miami-Dade Water and Sewer Authority. The

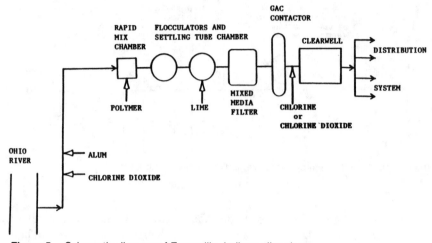

Figure 5. Schematic diagram of Evansville, Indiana pilot plant.

rated capacity of the plant was 2.63 m³/sec (60 mgd) and was operated near capacity. Raw water was supplied from seven wells, each approximately 27.4 m (90 ft) deep, to three upflow Hydrotreater® softeners where approximately 200 mg/L of lime was added. Activated silica (1.6 mg/L) was fed to the Hydrotreator influent. After recarbonation, 18 mg/L of chlorine was added to the Hydrotreator and routed to a chlorine contact basin with a retention time of 1.25 hr. Following chlorine contact, the water flowed to 12 rapid sand filters at a rate of 0.22 m³/sec–7.2 m/hr (5 mgd – 3 gpm/ft²) and then to the clearwell before distribution.

Bench-Scale Adsorption Unit

Flow diagrams for the bench-scale configurations used in the research effort are shown in Figures 7–9. Each glass column was 1.52 m (5 ft) long by 2.54 cm (1 in.) in internal diameter. A flow rate of 7.2 m/hr (3 gpm/ft²) was maintained by rotameters.

ORSANCO UTILITIES

Huntington, West Virginia

Raw source water from the Ohio River under normal operating conditions was coagulated by the addition of 8.6–34.2 mg/L of ferric sulfate and the pH was adjusted by 10.8–42.8 mg/L of lime (Figure 10).[7] No other chemical addition to the raw water was practiced except under unusual circumstances.

Three to five mg/L of chlorine was added before settling and 1.5–2.5

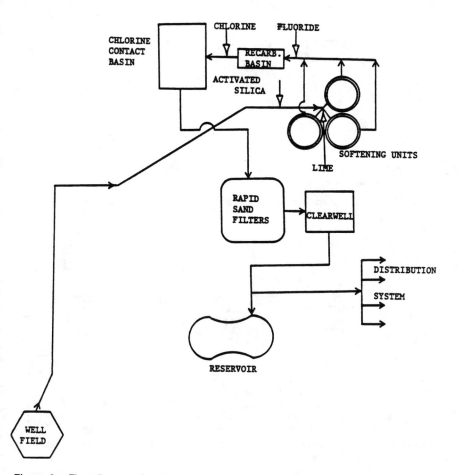

Figure 6.   Flow diagram of John E. Preston water treatment plant in Miami, Florida.

mg/L before discharge to the distribution system. Depending upon the normal fluoride concentration in the raw water (usually 0.1–0.3 mg/L), an additional dosage of 0.7–0.9 mg/L was added to obtain a 1-mg/L concentration.

Although no GAC pilot studies were conducted to select the optimum GAC or bed depth for organics control, Westvaco's WVW 14 × 40 GAC was evaluated based on its effective taste and odor control at the utility. Virgin GAC was used in a filtration/adsorption mode in a sand replacement bed. The bed was placed in service with 76 cm (30 in.) of GAC on top of 30 cm (12 in.) of sand and gravel. The mean hydraulic loading rate was 6.1 m/hr (2.6 gpm/ft²) and the mean EBCT was 7.2 min. Additional data were

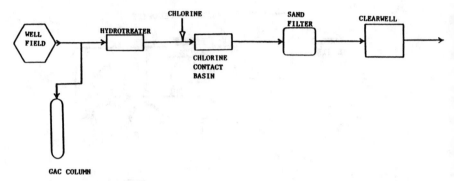

Figure 7. Flow diagram of raw water influent to bench-scale GAC adsorption columns in Miami, Florida.

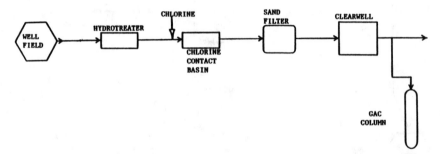

Figure 8. Flow diagram of clearwell influent to bench-scale GAC adsorption columns in Miami, Florida.

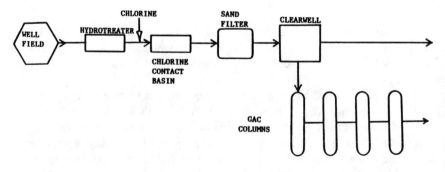

Figure 9. Flow diagram of bench-scale adsorption for deep bed GAC study in Miami, Florida.

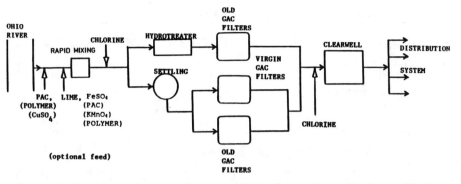

Figure 10. Treatment scheme for Huntington Water Corporation in Huntington, West Virginia.

collected on older (12- and 30-month-old) filter adsorber beds to evaluate performance over longer periods of operation.

Beaver Falls, Pennsylvania

Raw source water for the Beaver Falls water treatment plant was supplied by the Beaver River (Figure 11).[7] Primary coagulation of this water was accomplished by a liquid alum dose of about 21 mg/L. Occasionally, potassium permanganate was added at a dose of 2.0 mg/L when not going to breakpoint. Secondary coagulation and pH adjustment were accomplished

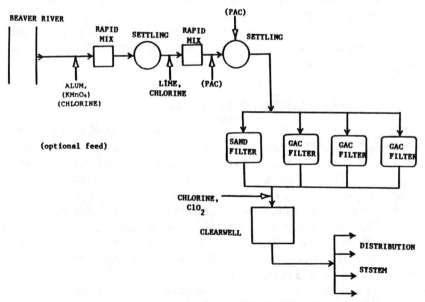

Figure 11. Water treatment scheme for Beaver Falls Authority, Pennsylvania.

Table 1. Hydraulic Data (Mean Value), Beaver Falls, Pennsylvania

	GAC	
Parameter	Filtrasorb 400	Hydrodarco 8 × 16
Loading rate, m/hr	3.1	3.1
(gpm/ft²)	(1.3)	(1.3)
Empty bed contact time, min	11.3	11.4
GAC depth, cm	62	62
(in.)	(24)	(24)
Sand and gravel depth, cm	30	30
(in.)	(12)	(12)

by about 25 mg/L of lime. Also added before secondary settling for break-point chlorination was about 9.0 mg/L of chlorine.

About 0.7 mg/L of fluoride was added to the secondary settled water to produce a final concentration of 1.0 mg/L. When used, about 14 mg/L of PAC was also added to the secondary settled water. Postfilter effluent (sand and GAC) was disinfected on the average with an applied dose of 0.5 mg/L chlorine and 0.1 mg/L of chlorine dioxide.

Two beds of virgin GAC were evaluated in the filtration/adsorption mode. A third bed contained Filtrasorb C®, a special Calgon product formulated for trihalomethane adsorption. One GAC bed contained 61 cm (24 in.) of Calgon's Filtrasorb 400 placed on top of 30 cm (12 in.) of sand and gravel. The second GAC bed contained 61 cm (24 in.) of ICI's Hydrodarco 8 × 16 placed on top of 30 cm (12 in.) of sand and gravel. The filter/adsorbers were geometrically identical except that the Calgon unit was designed with a tile bottom, while the ICI unit had a porous plate bottom. Hydraulic data for each unit are presented in Table 1.

PASSAIC VALLEY, LITTLE FALLS, NEW JERSEY

The Little Falls filtration plant source water was supplied by the Passaic River (Figure 12).[8] This plant employed prechlorination, alum coagulation, settling, filtration (anthracite coal over sand), and dechlorination by sulfur dioxide. During the study, an average of 28.2 mg/L of alum and 16.0 mg/L of chlorine were added prior to the raw water pumps. Sulfur dioxide at an average concentration of 0.9 mg/L was added after the settling basin prior to filtration. For pH control, sodium hydroxide (average of 27.6 mg/L) and postdisinfection chlorine (0.5 mg/L average) were added to the clearwell effluent. Average flow of the 378,500-m³/day (100-mgd)–designed plant was 208,000 m³/day (55 mgd).

A portion of the treated water (7192 m³/day or 1.9 mgd) that is normally distributed to the consumer was pumped to three cylindrical pressure GAC contactors. Each contactor was 8.5 m (28 ft) high and 2.4 m (8 ft) in

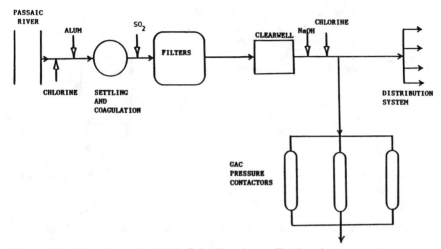

Figure 12. Flow schematic of Little Falls, New Jersey filtration plant.

diameter with an EBCT of approximately 8 min. GAC from three different manufacturers was evaluated simultaneously in the pressure contactors. The carbons consisted of ICI Hydrodarco 10 × 30, Westvaco WVW 12 × 40, and Calgon Filtrasorb 400.

THORNTON, COLORADO

Two water treatment plants were available to the city of Thornton for treating their drinking water (Figure 13).[9] One plant was the 56,780-m³/day (15-mgd) Thornton plant and the other one was the 75,700-m³/day (20-mgd) Columbine plant. The Columbine plant was the full-scale plant evaluated during the research project because of the contaminants in its raw water sources.

A major portion of Columbine's raw water was directly influenced by the Denver Metro Sewage Treatment Plant and various industrial dischargers. These discharges were into the South Platte River that infiltrated into various gravel lakes and wells located adjacent to the river. Columbine's raw water was drawn from these lakes and wells.

Potassium permanganate (0.4 to 0.6 mg/L) was added to the raw water for iron and manganese removal. For coagulation, ferric chloride was used at a concentration that varied from 5 to 20 mg/L. Breakpoint chlorination was accomplished by chlorine doses that at times ranged up to 50 mg/L. After premix, 2 to 4 mg/L of PAC was added for taste and odor control and some free chlorine removal. After PAC, lime was added for pH control to

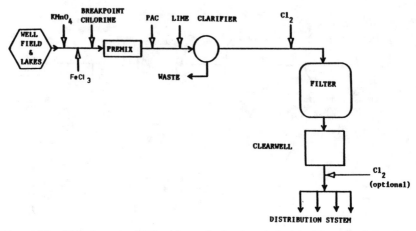

Figure 13.   Flow diagram of Columbine water treatment plant for Thornton, Colorado.

8. Following clarification, the water was chlorinated and filtered through eight dual-media filters.

Pilot Plant

A 10-gpm pilot plant consisting of a raw water rapid mix chamber, two-stage flocculation, settling basin, dual-media filter, and water storage chamber was used to simulate full-scale treatment (Figure 14). A portion of the raw water going to the full-scale Columbine plant was diverted to the pilot plant. Before filtration, a portion of the water was diverted to two GAC columns in series. After filtration, a portion of the water again was directed to two GAC columns in series. This provided the opportunity to evaluate combined filtration/adsorption and postfiltration adsorption.

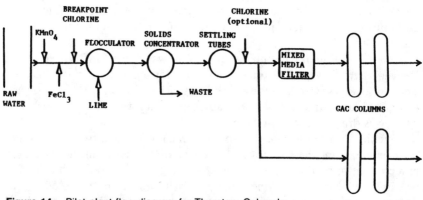

Figure 14.   Pilot plant flow diagram for Thornton, Colorado.

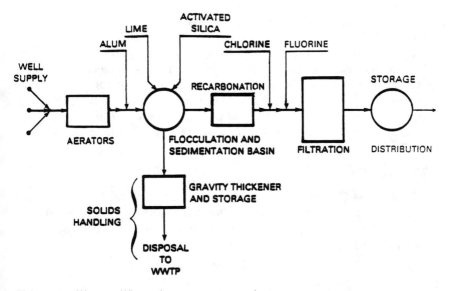

Figure 15. Wausau, Wisconsin water treatment plant.

Each of the four GAC columns was 15.2 cm (6 in.) in diameter and 2.4 m
(8 ft) tall with an effective depth of 1.7 m (5.5 ft) of Westvaco 12 × 40
GAC. The EBCT for each column was 10 min at a hydraulic loading rate of
9.6 m/hr (4 gpm/ft²) at a flow of 3.0 L/min (0.8 gpm).

WAUSAU, WISCONSIN

Several wells serve as the principal water source for the city of Wausau.[10]
These wells have a depth of about 30.5 m (100 ft) and a nominal capacity
range of 3.8–7.6 m³/min (1000–2000 gpm). Average daily flow from these
wells was about 0.2 m³/sec (4.4 mgd). Design peak flow was 0.5 m³/sec (12
mgd).

Water from the wells passed through two packed tower aeration units
(Figure 15) with design capacities of 5.7 m³/min (1500 gpm) and 7.6 m³/min
(2000 gpm). An air-to-water ratio of 30:1 was used to remove volatile
organic compounds from the water source. One unit was 2.4 m (8 ft) in
diameter while the other unit was 2.8 m (9.25 ft) in diameter and both
contained about 7.5 m (24.5 ft) of packing.

Alum was added prior to the clarifier at approximately 15 mg/L. In the
mixing zone of the clarifier, 45 mg/L of lime and 6 mg/L of activated silica
were added for iron and manganese removal. The clarifier was 21.3 m (70
ft) square with 4.7 m (15.5 ft) liquid depth. Clarified water flowed to a
recarbonation basin that was 4.0 m (13.2 ft) by 11.1 m (36.5 ft) with 3.0 m

(10 ft) liquid depth. Approximately 454 kg (0.5 tons) of CO_2 was used each day at a detention time of 10 min to produce a stable pH of 8.1. After recarbonation, 5 mg/L of liquid chlorine and 0.75 mg/L of sodium silicofluoride were added prior to filtration. Four gravity rapid sand filters with a total surface area of 288 m² (3100 ft²) were available for use. Average hydraulic loading based on 0.2 m²/sec (4.4 mgd) was 114 m/day (2800 gpd/ft²) using two filters. Capacity of the storage well was 0.04 m³/sec (1 mgd). Residuals in the distribution system were trace for chlorine and 1.1 mg/L for fluoride.

GAC Adsorber

A single GAC adsorber with a nominal EBCT of about 12 min was used to treat 0.378 m³/min (100 gpm). This adsorber was of cylindrical shell construction of 7.9 mm (5/16 in)–thick carbon steel designed to the following specifications.

diameter	2.2 m (7.0 ft)
vertical sidewall height	1.8 m (6.0 ft)
GAC depth	1.3 m (4.3 ft)
service flow	0.4 m³/min (98.3 gpm)
hydraulic loading	6.3 m/hr (2.6 gpm/ft²)
contact time	12.7 min
cross-sectional area	3.7 m² (38.5 ft²)
carbon mass	1991 kg (4390 lb)
carbon size	12 × 40 mesh

GAC Pilot Columns

The pilot system was composed of seven 0.05-m (2-in.) inside diameter glass columns that were configured for a series mode of operation. About 0.6 m (24 in.) of filter sand was placed in the first column to prevent any filterable particulate matter from depositing on the GAC columns during operation. Varying amounts of 12 × 40 mesh carbon were placed in the other glass columns with sampling ports located between each column. A loading rate of 4.7 m/hr (1.9 gpm/ft²) was used to simulate a gravity feed contactor system. All materials used in construction of the pilot plant were composed of either glass, stainless steel, or Teflon®. The amount of carbon used to produce various EBCTs is presented in Table 2 below.

PHILADELPHIA, PENNSYLVANIA

This project studied the unit operations of ozonation and GAC treatment when used in combinations of O_3/GAC, O_3/Sand, Cl_2/GAC, Cl_2-O_3/GAC, and no disinfection/GAC following the conventional treatment processes of raw water settling, coagulation, flocculation, sedimentation, and rapid sand filtration.[11] Figure 16 presents a schematic of the five advanced water

Table 2.   Volume of Carbon Used to Produce Specific Empty Bed Contact Times
(Wausau, Wisconsin)

Weight (lbs)	Dry Carbon (kg)	EBCT (min)
0.18	0.08	1.0
0.60	0.27	3.1
0.88	0.40	5.1
1.74	0.79	10.4
3.48	1.58	21.2
5.23	2.37	32.3

treatment (AWT) systems evaluated in this study. Two of these systems received their water from the chlorinated rapid sand filter effluent of the Torresdale Water Treatment Plant, a conventional coagulation/filtration plant that supplies the city of Philadelphia with half of its daily water requirements. The remaining three AWT systems received the nonchlorinated rapid sand filter effluent of a 30,000-gpd pilot plant. The primary

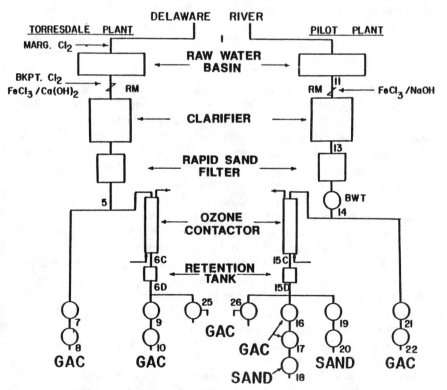

Figure 16.   Schematic of the five advanced water treatment systems evaluated at Philadelphia, Pennsylvania.

difference between the operation of the plants is that only the Torresdale plant chlorinates its process water; the pilot plant does not practice disinfection of any kind. The purpose of using the two plants was to determine the effects of chlorination and chlorine by-products on the AWT systems investigated. Following conventional treatment, the rapid sand filter effluents of the Torresdale and pilot plants were applied to the ozone and the carbon adsorbers.

Both ozone systems used co-current flow, in which both the ozone gas and water enter the contactors from the bottom. For the diffusion of the ozone gas into water, a porous dome diffuser was used. The adsorbers were constructed of 316-L stainless steel pipe that was 13 ft high and had an inside diameter of 10 in. To the extent possible, the two contacting systems were operated under identical conditions. Following ozonation, stainless steel retention tanks provided the ozonated water with sufficient contact time to yield an ozone residual below detectable limits. These tanks maximize the ozone/water contact time and minimize the possibility of oxidizing the carbon granules. A centrifugal pump system then transferred the water from the retention tanks to the O_3/GAC and O_3/sand systems.

Project Objectives

The study examined whether ozonation used as a pretreatment of GAC adsorption can increase the useful bed life of GAC sufficiently to justify its cost. The project was based on the understanding that ozone will transform some of the higher-molecular-weight humic substances into more readily biodegradable forms. These lower-molecular-weight organic compounds are then potentially available as a food source for microbes already present on the GAC bed. Preozonation may thus make available more adsorptive sites on the carbon for the less biodegradable, more harmful organic compounds that are poorly oxidized (e.g., chloroform and 1,2-dichloropropane). Preozonation may also help prolong GAC bed life by stripping volatile organic compounds from the process stream.

The project was conducted on both a pilot and laboratory scale to investigate the technical and economic feasibility of incorporating an ozonation and/or GAC unit process into a conventional water treatment train to remove trace organics of health concern. The relationship between adsorption and biological activity during water treatment with ozone and GAC was carefully evaluated, as were the effects of prechlorination on these mechanisms. The removal of trace organic substances of health concern at the ng/L level and the removal of TOC at the mg/L level were monitored along with microbial parameters of biological speciation and growth rates. The specific criteria chosen for GAC reactivation are listed in Table 3. Except for the criteria based on the cumulative breakthrough

Table 3. Organic Criteria for GAC Reactivation

Parameter	Criteria Investigated
DOC	Effluent levels of 0.5, 1.0, 1.5, and 2.0 mg/L
DOC, THMFP	50% cumulative breakthrough
Chloroform	Effluent levels of 1, 10, and 100 μg/L
1,2-dichloropropane	Effluent levels of 1 and 5 μg/L
All volatile organic compounds with average concentration > 1 μg/L	Average time to: initial breakthrough 100% breakthrough

of 50% of the dissolved organic carbon (DOC) and a trihalomethane formation potential (THMFP), all of the criteria investigated are based on the effluent of a single filter, and not on the average effluent of a multiple-filter operation.

The following questions were addressed in this project:

1. Can enhanced TOC removal before GAC treatment increase the capacity of the GAC for the trace organics of health concern?
2. Can ozonation as a pretreatment to GAC adsorption increase the useful bed life of GAC sufficiently to justiy its cost?
3. How does prechlorination affect the ozonation process and the GAC adsorption capacity for chlorinated organics and other volatile organics?

ORGANIC PARAMETERS ANALYZED

Although all of the field projects had a common objective to evaluate the effectiveness of GAC for removal of organic contaminants from drinking water, they were somewhat different in their approach. Also, in most cases, different sources of raw water were evaluated. Therefore, the same constituents were not present at all locations.

Sampling generally consisted of the raw source water, coagulated/settled water, sand filter effluent, and GAC effluent. Listed in Table 4 is a combination of all of the different organic compounds that were in the analytical scheme for the projects. Not all of these compounds were detected. In addition to these specific compounds, general parameters such as TOC and total organic halogen (TOX) were analyzed.

Table 4.   Specific Organics Evaluated During GAC Projects

Compound	Location*	Compound	Location*
Acenaphthene	MI, JP, EV	Para, para'-DDE	JP
Acenaphthylene	EV	DDT	O
Acetonaphthone-2'	JP	Ortho, para'-DDT	JP, PA
Acetone	PH	Para, para'-DDT	EV, JP, PA
Acrolein	EV	Decanal	CW
Acrylonitrile	EV	Decane-N	JP
Aldrin	CW, JP, EV, MI, PA, O	Dibenzo(a,h) anthracene	PA
Aniline	CW	Dibromoethane-1,2	PA
Anthracene	EV, JP	Dibromomethane	CW, EV, JP
Atrazine	JP	Di N-butyl phthalate	EV, MI, PA
Benzene	CW, EV, JP, MI, PH	Dichlorobenzene, summation of -m, -o, -p	MI
Benzidine	CW, EV	Dichlorobenzene-meta	EV, JP, MI, O
Benzo(a)anthracene	EV	Dichlorobenzene-ortho	CW, EV, MI, O
Benzo(k)fluoranthrene	EV	Dichlorobenzene-para	EV, MI, CW, O
Benzofluoranthrene-3,4	EV	Dichlorobenzene-1,2	PH
Benzonitrile	JP	Dichlorobenzidine-3,3	CW, EV
Benzo(g,h,i)perylene	EV	Dichlorodifluoromethane	EV
Benzo(a)pyrene	EV	Dichloroethane-1,1	O
Benzyl cyanide	JP	Dichloroethane-1,2	CW, EV, JP, MI, PA, O, PH
Biphenyl	JP	Dichloroethene-1,1	EV, MI, PA
Bis(2-chloroethyl) ether	EV, O, PH	Dichloroethene-cis-1,2	EV, MI, JP, W
Bis(2-chloroisopropyl) ether	EV, O	Dichloromethane	CW, JP
Bis(2-chloroethoxy) methane	EV	Dichlorophenol-2,3	EV
Bis(2-ethylhexyl) phthalate	CW, EV, JP	Dichloropropane-1,2	EV, O, PH
Bromochloromethane	JP	Dichloropropene-1,3,cis	O
Bromophenyl ether, 4	EV, O	Dichloropropene-1,3,trans	O
Butyl benzyl phthalate	CW, EV	Dichloropropylene-1,3	EV
N-Butyl isobutyl phthalate	CW	Dieldrin	EV, JP, MI, PA, O
Carbon tetrachloride	CW, EV, JP, MI, PA, O, PH	Diethyl phthalate	CW, EV, JP
Alpha chlordane	JP	Diiodomethane	PA
Chlordane	EV, MI	Di-Isobutyl phthalate	CW, JP, MI
Gamma chlordane	JP, PA	Dimethylbenzene-1,3 & -1,4	CW
Alpha chlordene	JP	Dimethylbenzene-1,2	CW
Beta chlordene	JP	Dimethyl phenol-2,4	EV
Gamma chlordene	JP	Dimethyl phthalate	CW, EV, JP
Chlorobenzene	CW, EV, JP, MI, O, PH	Dimethyl pyrene-2,4	CW
Chloro-m-cresol, P	EV	Dinitro-o-cresol-4,6	EV
Chloroethane	EV	Dinitrophenol-2,4	EV
Chlorophenyl phenyl ether-4	O	Dinitrotoluene-2,4	CW, EV
Dichloroethene-trans-1,2	PA	Dinitrotoluene-2,6	CW, EV
Chloronaphthalene-2	O	Di-noctyl phthalate	CW
1-Chloro-2-nitrobenzene	JP	Diphenylhydrazine-1,2	CW, EV
Chlorotoluene-P	MI	Dipropyl phthalate	JP
Dimethyl-1,2,3,5,6- tetrachlorotere phthalate	JP	Dodecane-N	JP
		Eicosane-N	JP
DDD	O	Endosulfan-1	JP
Ortho, para'-DDD	JP	Endosulfan-2	JP
Meta, para'-DDD	JP	Endrin	EV, JP, PA
Para, para'-DDD	JP	Ethyl benzene	CW, EV, JP, W, PH
DDE	O	Ethyl benzoate	EV, MI
Ortho, para'-DDE	JP, PA	Fluoranthene	EV, JP, MI
		Fluorene	EV, JP, MI
		Heptachlor	MI, PA, O

Table 4.   (contd.)

Compound	Location*	Compound	Location*
Heptachlor epoxide	EV, MI, PA, O	Polychlorinated biphenyl	PA
Heptadecane-N	JP	1248	
Hexachlorobenzene	EV, JP, MI	Polychlorinated biphenyl	MI, PA, MI
Hexachlorobutadiene	EV, O	1254	
Hexachlorocyclohexane	EV, JP, MI, PA, O	Pentachloronitrobenzene	EV, JP, MI
(alpha benzene		Pentachlorophenol	EV
hexachloride)		Pentadecane-N	JP
Hexachlorocyclohexane	EV, JP, PA, O	Perthane	JP
(beta benzene		Phenanthrene	EV, JP, MI
hexachloride)		Phenol	EV, JP
Hexachlorocyclohexane	EV, O	Photodieldrin	JP
(delta benzene		Propylbenzene	CW
hexachloride)		Pyrene	EV, JP, MI
Hexachlorocyclohexane	EV, JP, MI, O	Pyridine	CW
(gamma benzene		Tetrachlorodibenzodioxin	EV
hexachloride)—lindane		Tetrachloroethane-1,1,2,2	CW, EV, JP, PA
Hexachlorocyclopentadiene	EV	Tetrachloroethene	EV, JP, MI, PA,
Hexachloroethane	CW, EV, O	(also tetrachloroethylene)	W, PH
Hexadecane-N	JP	Tetrachloromethane	JP
Hexane	CW	Tetradecane-N	JP
Indeno (1,2,3-C,D) pyrene	EV	Tetralin	CW
Isophorone	CW, EV, JP	Tetramethylbenzene-1,2,3,5	CW
Methoxychlor	EV, JP, PA	Toluene	CW, EV, JP, MI,
Methylbenzene-1-ethyl-2	CW		W, PH
Methylbenzene-1-ethyl-4	CW	Toxaphene	EV, PA
Methyl bromide	EV	Trichlorobenzene	CW, EV, PA, O
Methyl chloride	EV, MI	(1,3,5,-1,2,4,-1,2,3)	
Methylene chloride	EV, MI, PA, PH	Trichloroethane-1,1,1	CW, EV, MI, PA,
Methylnaphthalene	CW, JP	(also methyl chloroform)	O, W, PH
Mirex	JP, PA	Trichloroethane-1,1,2	JP, PA
Naphthalene	CW, EV, JP, MI	Trichloroethene	CW, EV, JP, MI, P,
Nitrobenzene	CW, EV, JP	(also trichloroethylene)	W, PH
Nitrophenol-2	EV	Trichlorofluoromethane	EV
Nitrophenol-4	EV	Trichlorophenol-2,4,6	EV
N-Nitroso di-n-propylamine	EV	Trichlorophenoxy acetic	EV
N-Nitroso dimethylamine	EV	acid-2,4,5	
N-Nitroso diphenylamine	EV	Trichlorophenoxy acetic	EV
Nitrotoluene-2	EV, JP, MI	acid-2,5,5	
Nonadecane-N	JP	Tridecane-N	JP
Nonanal	CW	Trimethylbenzene-1,2,4	CW
Octadecane-N	JP	Trimethyl-1-pentene-2,2,4	CW
Oxychlordan	EV, JP, MI	Triphenylmethane	JP
Polychlorinated biphenyl	PA	Undecanal	CW
1016		Undecane-N	JP
Polychlorinated biphenyl	PA	Vinyl chloride	MI, PA, JP, W
1242		Xylene	CW, EV, W, PH

*CW—Cincinnati Water Works.   O—ORSANCO (Huntington, Beaver Falls).
 EV—Evansville.   PA—Passaic.
 JP—Jefferson Parish (I).   TH—Thornton.
 MA—Manchester.   W—Wausau.
 MI—Miami (I).   PH—Philadelphia.

REFERENCES

1. Miller, R., and D. J. Hartman. "Feasibility Study of Granular Activated Carbon Adsorption and On-Site Regeneration," Volume 1, U.S. EPA Report 600/2-82-087A (October 1982).
2. "Recommended Standards for Water Works," in *Policies for the Review and Approval of Plans and Specifications for Public Water Supplies* (Albany, NY: Health Education Service, 1982).
3. Kittredge, D., R. Beaurivage, and D. Paris. "Granular Activated Carbon Adsorption and Fluid-Bed Reactivation at Manchester, New Hampshire," U.S. EPA Report 600/2-83-104 (March 1984).
4. Koffskey, W. E., N. V. Brodtmann, and B. W. Lykins, Jr. "Organic Contaminant Removal in Lower Mississippi River Drinking Water by Granular Activated Carbon Adsorption," U.S. EPA Report 600/2-83-032 (June 1983).
5. Lykins, Jr., B. W., M. Griese, and D. B. Mills. "Chlorine Dioxide Disinfection and Granular Activated Carbon Adsorption," U.S. EPA Report 600/2-82-051 (May 1983).
6. Wood, P. R., and D. F. Jackson. "Removing Potential Organic Carcinogens and Precursors from Drinking Water," U.S. EPA Report 600/2-80-130a (August 1980).
7. "Water Treatment Process Modifications for Trihalomethane Control and Organic Substances in the Ohio River," Ohio River Valley Water Sanitation Commission, U.S. EPA Report 600/2-80-028 (March 1980).
8. Inhoffer, W. R. "Evaluating the Use of Granular Activated Carbon at Passaic Valley, New Jersey," U.S. EPA Report 600/2-86-028 (October 1986).
9. Speed, M. A., A. Bernard, R. P. Arber, G. C. Budd, and F. J. Johns. "Treatment Alternatives for Controlling Chlorinated Organic Contaminants in Drinking Water," U.S. EPA Report 600/52-87-011 (April 1987).
10. Hand, D. W., J. C. Crittenden, J. M. Miller, and J. L. Gehin. "Performance of Air Stripping and GAC for SOC and VOC Removal from Groundwater," Cooperative Agreement CR-11150, U.S. EPA (September 1988).
11. Neukrug, H. M., M. G. Smith, J. T. Coyle, J. P. Santo, J. McElhaney, I. H. Suffit, S. W. Maloney, P. C. Chrostowski, W. Pipes, J. Gibs, and K. Bancroft. *Removing Organics from Philadelphia Drinking Water by Combined Ozonation and Adsorption* (Cincinnati, OH: U.S. EPA Office of Research and Development, June 1983).

CHAPTER 4

Comparative Analysis of Field-Scale Projects

INTRODUCTION

One of the major purposes of the field-scale projects was to conduct a comparative analysis among various modes of GAC operation. In addition, the studies were designed to evaluate the effect of variables such as empty bed contact time and GAC type on the performance of the various systems studied. The following sections discuss these issues.

ADSORBER VS FILTER ADSORBER

Options for GAC adsorption include building separate contactors after sand filtration (adsorber) and replacing the filter media in the filter shell (filter-adsorber). Each of these options has some advantages and certainly, if performance is equivalent, the filter adsorber is the least expensive of the two options.

In three field-scale locations (Cincinnati, Ohio, Jefferson Parish, Louisiana, and Thornton, Colorado) both the adsorber and filter-adsorber modes of operation were used. The adsorber mode consisted of sand filtration prior to GAC adsorption and the filter-adsorber mode consisted of applying coagulated and settled effluent directly to the GAC. At Cincinnati, the filter adsorber (sand replacement) EBCT was 7.5 min; therefore, the EBCT for the adsorber was established as 7.2 min for comparison. The actual TOC concentrations applied and removed for these contact times are shown in Figure 1. When variables such as flow rate, influent concentration, and amount of GAC are considered, as shown in Figure 2, the apparently

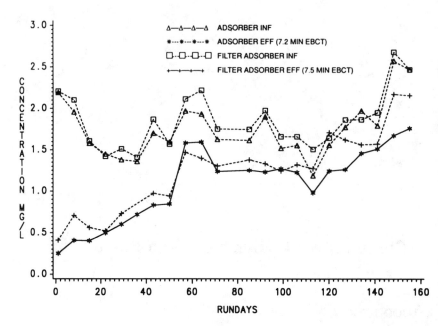

Figure 1.   Comparison of adsorber and filter adsorber performance for TOC—Cincinnati, Ohio.

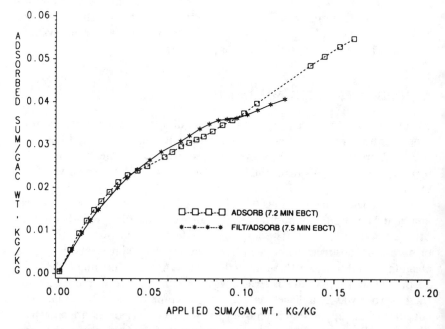

Figure 2.   TOC loading on virgin GAC for adsorber and filter adsorber at comparable EBCTs—Cincinnati, Ohio.

slightly better performance of the adsorber can probably be attributed to a slightly lower influent concentration.

At Jefferson Parish, a comparison was made of the adsorber and filter adsorber modes of operation for all four phases of the project. As with any full-scale project, maintenance of exact operating conditions for comparative purposes is difficult. There were, however, two phases (IIB and III) in which TOC influent concentrations and EBCTs were similar enough for comparison of the adsorber and filter adsorber performance. Phase IIB showed better performance for the adsorber, but its EBCT was 3.3 min longer than the filter adsorber (17.0 vs 13.7 min). The TOC performance of the adsorber and filter adsorber was most comparable in Phase III. A slight advantage was seen with the filter adsorber and was probably due to a 4.4-min-longer EBCT (23.2 vs 18.8 min). Figure 3 shows the percent TOC removal for those two phases.

Consistent detection of a few specific organics also showed in Phase IIB and III that the adsorber and filter adsorber performed essentially the same. For instance, 1,2-dichloroethane and trichloroethylene followed each other except for an occasional discrepancy. This closeness of performance occurred even during desorption of the 1,2-dichloroethane (Figures 4 and 5).

By using the pilot plant at Thornton for evaluation of the adsorber and filter adsorber GAC schemes, a consistent EBCT could be maintained for both. With an EBCT of 20 min for each adsorption system, one can see from the average concentrations presented in Figure 6 that for all modes evaluated, the filter adsorber effluent concentration for TOC was lower than for the adsorber effluent. This occurred although the average applied concentration was similar.

For terminal TOX, as shown in Figure 7, longer runs (six months in this case for Modes 1 and 4) may be accomplished by applying clarifier effluent directly to GAC for removal of TOX precursors as indicated by the lower terminal values. The average instantaneous TOX (Figure 8) in Mode 4 adsorber effluent was lower than the filter adsorber effluent. This probably occurred because chloramines were produced in the treatment process for disinfection and only partially reacted with the TOX precursors.

PILOT VS FULL-SCALE *GAC* SYSTEMS

If GAC is to be considered as a treatment option by water utilities in the United States it is very important to be able to predict its performance using pilot-scale facilities. The ability to predict full-scale GAC performance from pilot systems provides utilities with an economical screening mechanism to utilize before making decisions relative to GAC use. In Cincinnati, the 7.6-cm (3 in.) diameter by 76.2-cm (30 in.) depth columns had approxi-

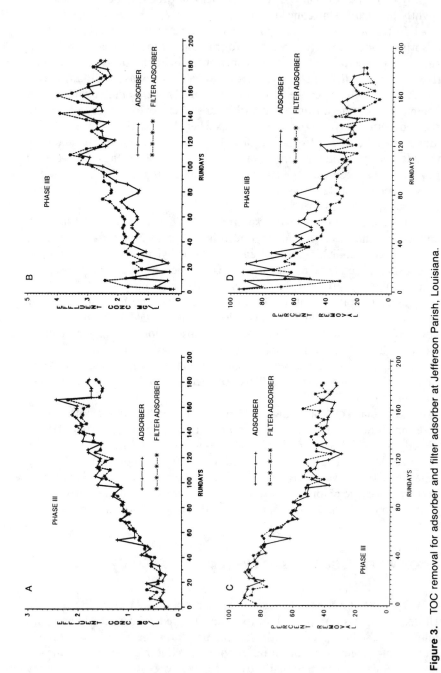

Figure 3. TOC removal for adsorber and filter adsorber at Jefferson Parish, Louisiana.

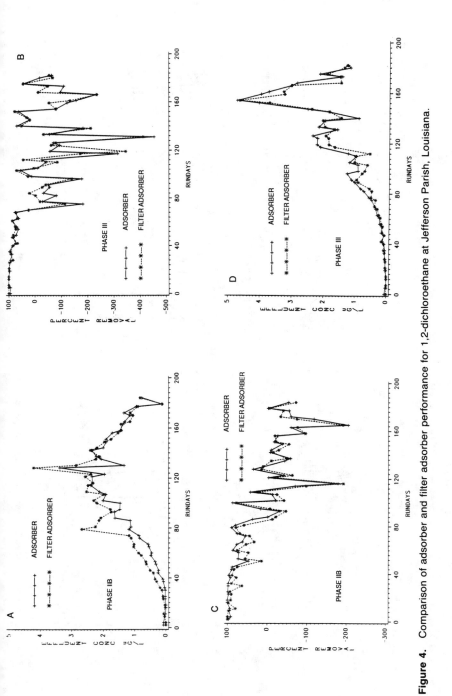

Figure 4. Comparison of adsorber and filter adsorber performance for 1,2-dichloroethane at Jefferson Parish, Louisiana.

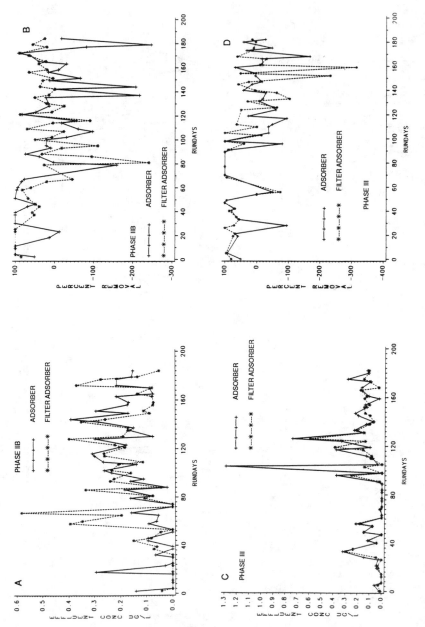

Figure 5. Comparison of adsorber and filter adsorber performance for trichloroethylene at Jefferson Parish, Louisiana.

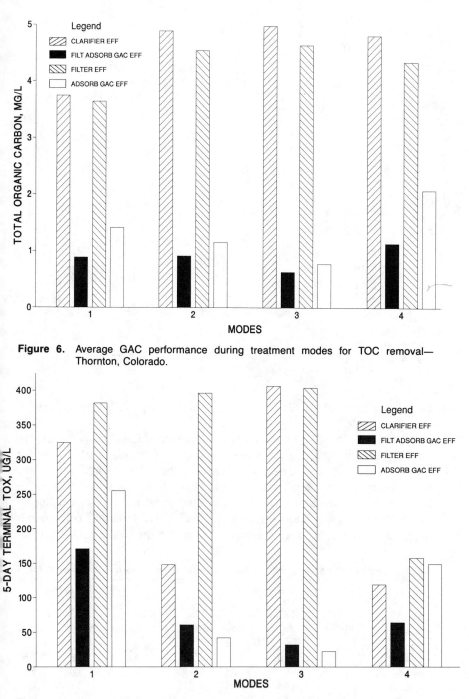

Figure 6. Average GAC performance during treatment modes for TOC removal—Thornton, Colorado.

Figure 7. Average GAC performance during treatment modes for TOX removal—Thornton, Colorado.

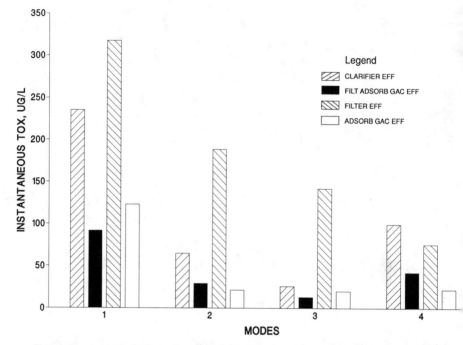

Figure 8. Average GAC performance during treatment modes for instantaneous TOX removal—Thornton, Colorado.

mately the same hydraulic loading applied as in the full-scale systems as shown in Table 1.

In Phases 3–0 and 3–1 (virgin and once-reactivated GAC, respectively) the pilot system predicted the full-scale adsorber performance within 10% for up to 100 days in most cases. In Phase 3–2 (twice-reactivated GAC), the pilot system prediction of full-scale adsorber performance was within 5% for 60 days of operation (Figure 9). For all phases, the full-scale adsorber data showed a slightly more efficient removal of TOC than the corresponding pilot column. The performance differences in the two systems have not

Table 1. Hydraulic Loading Rate of Pilot and Full-Scale Systems—Cincinnati, Ohio

GAC Filter System	Phase	Hydraulic Loading Adsorber			Hydraulic Loading Filter Adsorber		
		m³/hr/m²	gpm/ft²	EBCT	m³/hr/m²	gpm/ft²	EBCT
Pilot	3–0	18.2	7.6	14.8	6.0	2.5	7.6
Full-scale		17.8	7.4	15.3	6.0	2.5	7.5
Pilot	3–1	17.0	7.1	15.8	6.0	2.5	7.6
Full-scale		17.8	7.4	15.3	6.0	2.5	7.5
Pilot	3–2	16.6	6.9	16.4	5.8	2.4	7.8
Full-scale		17.8	7.4	15.3	6.0	2.5	7.5

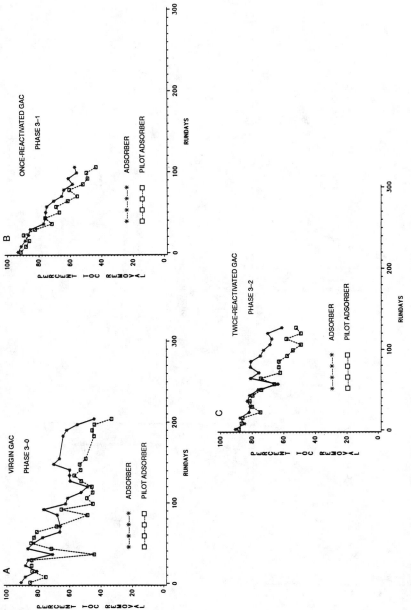

Figure 9. Prediction of adsorber performance for TOC using pilot simulation at Cincinnati, Ohio—Phase 3–2.

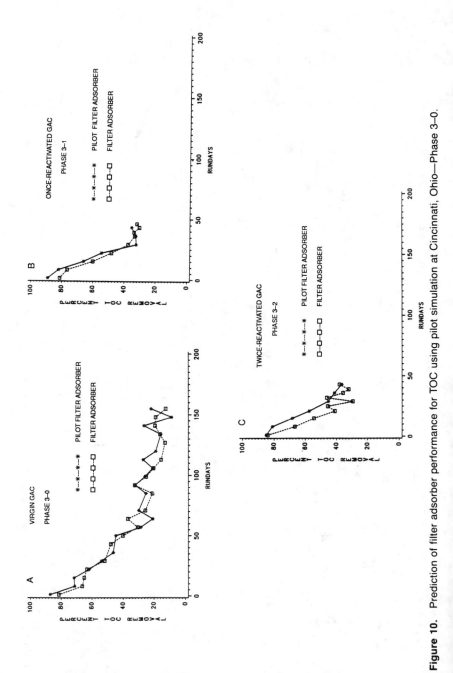

Figure 10. Prediction of filter adsorber performance for TOC using pilot simulation at Cincinnati, Ohio—Phase 3–0.

Table 2.   Average EBCT of Pilot and Full-Scale Systems—Jefferson Parish, Louisiana

	Phase IIB			Phase III		
	Hydraulic Loading			Hydraulic Loading		
	m³/hr/m²	gpm/ft²	EBCT	m³/hr/m²	gpm/ft²	EBCT
Full-scale adsorber	2.6	1.07	17.0	2.3	0.95	18.8
Pilot adsorber	2.4	1.00	21.2	2.2	0.90	21.7
Full-scale filter adsorber	2.8	1.16	13.7	1.8	0.75	23.2
Pilot filter adsorber	2.7	1.13	16.8	2.2	0.90	20.7

been completely explained, but maintenance of consistent and accurate flows throughout the testing period is suspect.

Essentially identical TOC removals for the pilot and full-scale filter adsorber systems were seen in Phase 3–0 (Figure 10). In the other two runs (3–1 and 3–2), the pilot predictor in most cases removed more TOC than the full-scale sand replacement filter.

At Jefferson Parish, similar predictive performance was also noted with the pilot column simulators of the full-scale systems. As with the Cincinnati study, measurement and control of flow and GAC in both systems to maintain an equal situation was difficult. As shown in Table 2, the EBCTs were somewhat different between the pilot and full-scale systems although the hydraulic loadings were similar.

This difference in EBCT is also reflective of the comparative performance difference between the full-scale systems and their respective pilot simulators. For Phase IIB, the pilot adsorber removed on the average 8% more TOC than the full-scale adsorber and in Phase III, 7% more (Figures 11 and 12). The pilot filter adsorber performance for TOC during Phase IIB showed an average 13% higher removal than did the sand replacement bed (Figure 13). Better pilot simulator performance was expected because of a longer EBCT, but in Phase III the pilot simulator removed on the average 6% more TOC than the corresponding full-scale filter adsorber, although the EBCT was 2.5 min less (Figure 14). As mentioned previously, flow rate measurements at times were difficult and can be a possible explanation for this discrepancy.

Examination of 1,2-dichloroethane, the most consistent specific organic identified at Jefferson Parish other than the trihalomethane species, showed similar pilot column predictive capabilities. For instance, in Phase IIB the pilot column simulator for both the adsorber and filter adsorber showed similar effluent concentrations to their full-scale counterparts (Figures 15 and 16). In Phase III, both the adsorber and the filter adsorber pilot column simulators exhibited somewhat greater adsorption efficiencies than the full-scale systems (Figures 17 and 18).

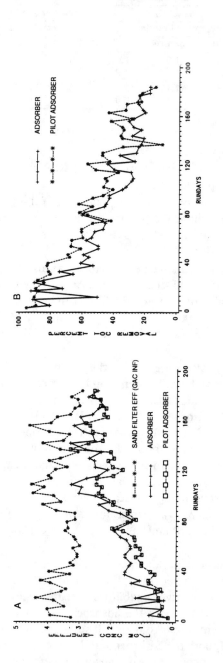

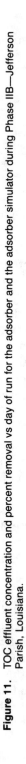

Figure 11. TOC effluent concentration and percent removal vs day of run for the adsorber and the adsorber simulator during Phase IIB—Jefferson Parish, Louisiana.

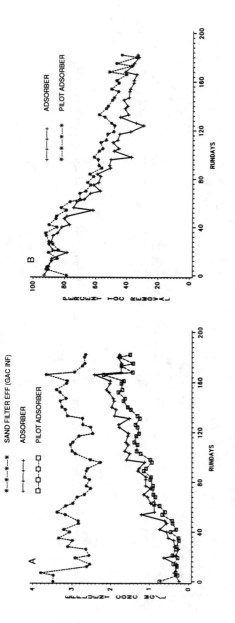

Figure 12. TOC removal for the adsorber and the pilot adsorber during Phase III—Jefferson Parish, Louisiana.

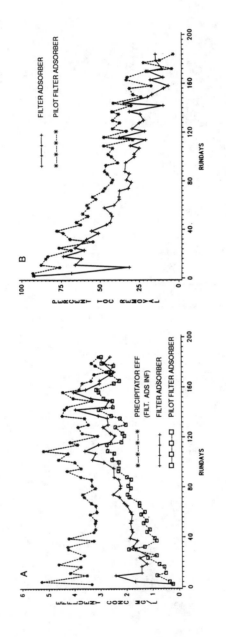

Figure 13. TOC removal for the filter adsorber and the filter adsorber simulator during Phase IIB—Jefferson Parish, Louisiana.

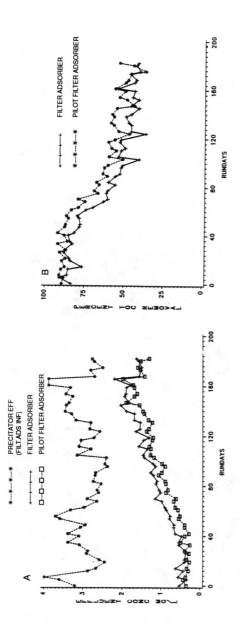

Figure 14. TOC removal for the filter adsorber and the pilot filter adsorber during Phase III—Jefferson Parish, Louisiana.

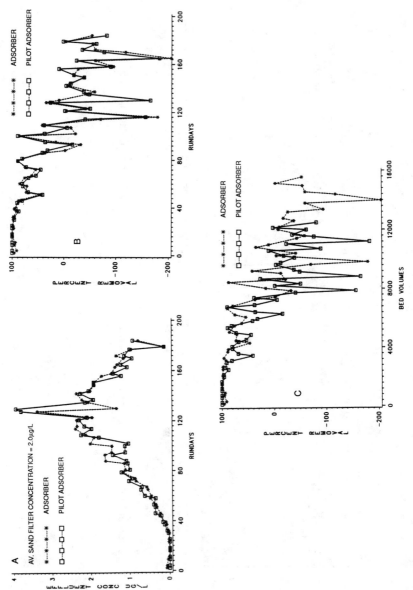

Figure 15. Removal of 1,2-dichloroethane through the adsorber and pilot adsorber during Phase IIB—Jefferson Parish, Louisiana.

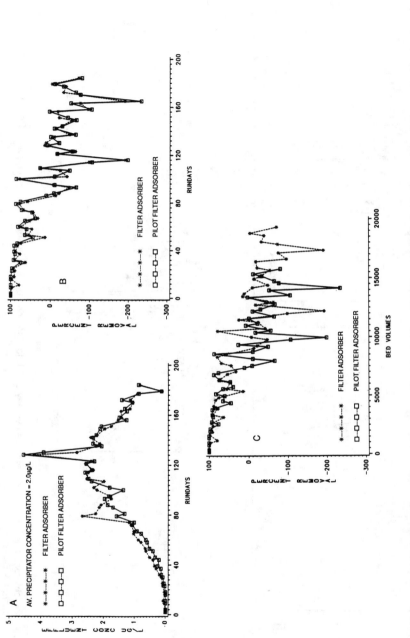

Figure 16. Removal of 1,2-dichloroethane through the filter adsorber and the pilot filter adsorber during Phase IIB—Jefferson Parish, Louisiana.

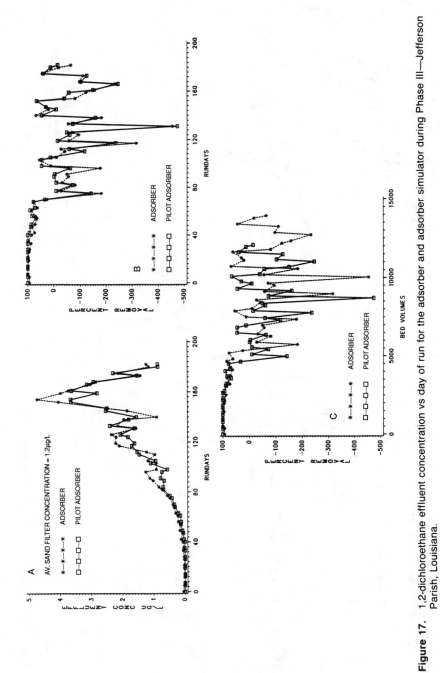

Figure 17. 1,2-dichloroethane effluent concentration vs day of run for the adsorber and adsorber simulator during Phase III—Jefferson Parish, Louisiana.

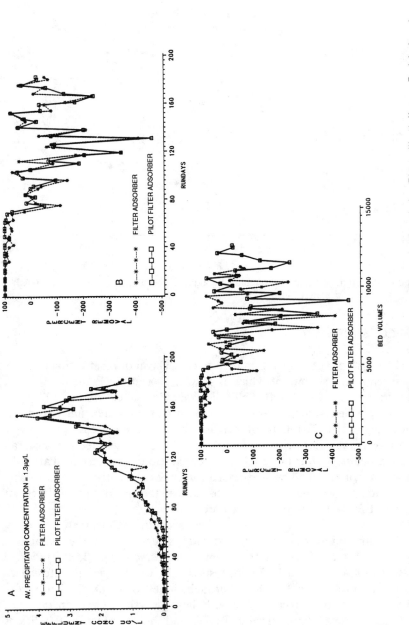

Figure 18. Removal of 1,2-dichloroethane through the filter adsorber and pilot filter adsorber during Phase III—Jefferson Parish, Lousiana.

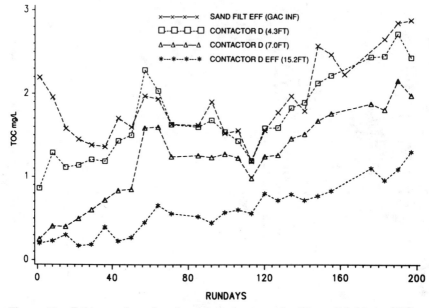

Figure 19. Total organic carbon breakthrough curves for Phase 3-0 (virgin GAC)—Cincinnati, Ohio.

EFFECTS OF *EBCT*

The length of GAC operation before replacement or reactivation depends on several factors, one of which is EBCT. If a utility is required to use existing filters, very little flexibility is available for selection of EBCT. In designing a new system, however, the utility has an opportunity to determine the best EBCT relative to performance criteria and cost.

A GAC exhaustion criterion of 1.0 mg/L TOC was selected at Cincinnati as a reasonable performance standard. Exhaustion was determined after the 1.0-mg/L criterion had been exceeded for three consecutive sampling days. This criterion will be used below to discuss the performance effects of varying EBCTs in lieu of a better criterion.

As shown previously, the GAC contactor depth at Cincinnati was 4.6 m (15 ft) with sampling ports located at various depths within the contactors. Samples taken at 1.3 m (4.3 ft), 2.1 m (7.0 ft), and 4.6 m (15.0 ft) yielded EBCTs of 4.4, 7.2, and 15.2 min, respectively. By applying the aforementioned criterion to the TOC breakthrough curves for virgin GAC in Figure 19, one can see the effects of EBCT. Relating length of GAC operation to carbon use rate as summarized in Table 3 and shown in Figure 20 indicates that longer EBCTs provided more efficient use of the GAC.

At Jefferson Parish, pilot columns were used to evaluate the effects of EBCT. By applying the same TOC exhaustion criterion (1.0 mg/L) to the four pilot series columns as shown in Figure 21, one can see a longer

Table 3. Summary Data at TOC Exhaustion of 1.0 mg/L Using Virgin WVG 12 × 40 GAC—Cincinnati, Ohio

GAC Depth	EBCT (min)	Hydraulic Loading m³/hr/m²	Hydraulic Loading gpm/ft²	TOC Exhaustion Time (days)	Carbon Use Rate (kg/mil L)	Organic Loading (g/kg)
1.3 m (4.3 ft)	4.4	17.8	7.4	22	46	15
2.1 m (7.0 ft)	7.2	17.8	7.4	71	34	25
4.6 m (15.0 ft)	15.2	17.8	7.4	204	26	51

operational time with increased EBCT through 31.4 min. No apparent advantage was gained by additional contact time above that EBCT, although some improvement in GAC effluent quality was noted. Essentially, a proportional increase in operation time before exhaustion relative to EBCT was noted through 31.4 min (Table 4).

The above TOC data for Jefferson Parish were produced using the same type of GAC that was used at Cincinnati (WVG 12 × 40). A different GAC (Filtrasorb 400, or F400) was also evaluated at a different time period with a lower average applied TOC (3.6 mg/L vs 2.9 mg/L). Consequently, the days of operation for the Filtrasorb 400 before exhaustion were longer (Figure 22) although the EBCTs were comparable (Tables 4 and 5).

In Miami, Florida, a series of pilot columns was used to produce an ultimate EBCT of 24.8 min at a hydraulic loading of 7.2 m/hr (3 gpm/ft²). The finished water TOC concentration at Miami was generally in the area of

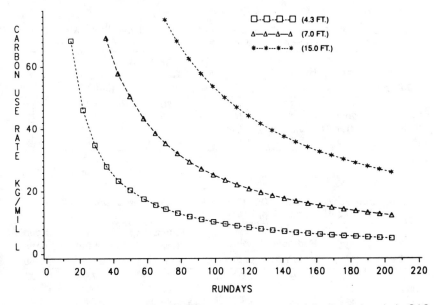

Figure 20. Carbon use rate for TOC at various contactor bed depths during virgin GAC evaluation at Cincinnati, Ohio.

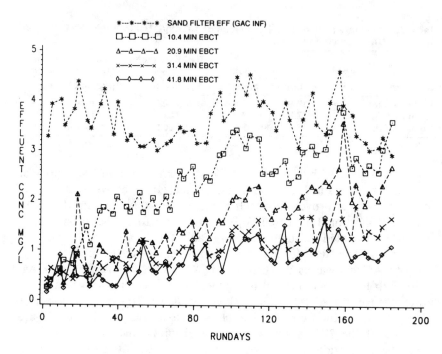

Figure 21. Comparison of TOC removal for various contact times using adsorbers with WVG 12 × 40 GAC—Jefferson Parish, Louisiana.

Table 4. Summary Data at TOC Exhaustion of 1.0 mg/L Using Virgin WVG 12 × 40 GAC—Jefferson Parish, Louisiana

GAC Depth	EBCT (min)	Hydraulic Loading m³/hr/m²	Hydraulic Loading gpm/ft²	TOC Exhaustion Time (days)	Carbon Use Rate (kg/mil L)	Organic Loading (g/kg)
36 in.	10.4	5.3	2.16	30	94	31
72 in.	20.9	5.3	2.16	79	36	70
108 in.	31.4	5.3	2.16	106	27	103
144 in.	41.8	5.3	2.16	106	27	109

Table 5. Summary Data at TOC Exhaustion of 1.0 mg/L Using Virgin F400 GAC—Jefferson Parish, Louisiana

GAC Depth	EBCT (min)	Hydraulic Loading m³/hr/m²	Hydraulic Loading gpm/ft²	TOC Exhaustion Time (days)	Carbon Use Rate (kg/mil L)	Organic Loading (g/kg)
36 in.	11.6	4.9	2.02	42	73	32
72 in.	23.2	4.9	2.02	105	30	78
108 in.	34.7	4.9	2.02	140	22	107
144 in.	46.3	4.9	2.02	159	20	126

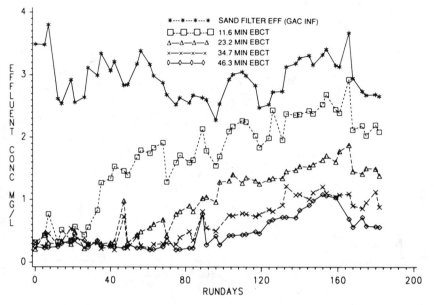

Figure 22. Comparative TOC removals for various contact times using adsorbers with F400 GAC—Jefferson Parish, Louisiana.

5 mg/L, compared to Cincinnati and Jefferson Parish ranges of 2 to 4 mg/L. Using the same GAC exhaustion criterion (1.0 mg/L TOC), indications are that at Miami, length of operation before exhaustion is less than was seen at Cincinnati and Jefferson Parish EBCTs (Figure 23 and Table 6).

If the GAC criterion, however, were selected at Miami when a plateau occurs in the breakthrough curve (steady state) then the operational time could be extended significantly. For instance, at 24.8 min EBCT the 1.0-mg/L TOC criterion would allow 35 days before exhaustion and the plateau method 60 days. The performance loss by allowing the GAC adsorber to continue operating until steady state at Miami would be an average increase in effluent concentration of 2.2 mg/L at 24.8 min EBCT (total TOC effluent concentration was 3.2 mg/L).

During the pilot plant study at Wausau, Wisconsin the TOC influent concentration averaged about 8 mg/L. Even at an EBCT of 32 min, the 1.0-mg/L TOC criterion was exceeded after 15 days of operation. Steady-state conditions appeared to start occurring around 30 days of operation for 10 min EBCT and 150 days for 30 min EBCT at a TOC of about 6 mg/L. Throughout the rest of the one-year study, the TOC concentration of the GAC effluent gradually increased until it approximated the influent concentration. In Jefferson Parish 1,2-dichloroethane was detected consistently enough to evaluate. When using WVG 12 × 40 GAC, saturation, as determined by 0% removal, was 80, 110, and 160 days for 10.4, 20.9, and 31.4 min EBCT, respectively. For 41.8 min EBCT, saturation had not

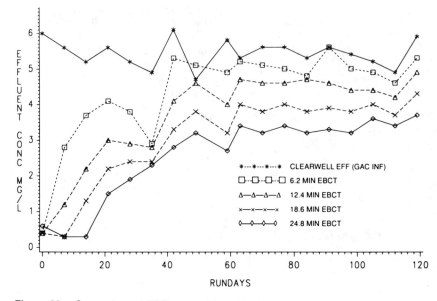

Figure 23. Comparison of TOC removal for various contact times using adsorbers with F400 GAC—Miami, Florida.

occurred after 180 days of operation. Desorption occurred at varying degrees for each EBCT that experienced saturation (Figure 24).

A somewhat different time of saturation was noted for the F400 GAC as shown in Figure 24. After 11.6, 23.2, 34.7, and 46.3 min EBCT, saturation occurred in 70, 110, 130, and 160 days, respectively. This GAC seemed to perform similarly to the WVG up to about 23 min EBCT, but after that the WVG removed the 1,2-dichloroethane more effectively even at a higher average GAC influent concentration (2.0 vs 1.2 µg/L). This performance difference is also illustrated by mass loading (Figure 25).

In Miami, the average GAC influent concentration of 1,2-dichloroethane was 19.9 µg/L, considerably higher than at Jefferson Parish. Although this was the case, the length of GAC operation before saturation was much longer than at Jefferson Parish, indicating that competitive adsorption was

Table 6. Summary Data at TOC Exhaustion of 1.0 mg/L Using F400 GAC—Miami, Florida

GAC Depth	EBCT (min)	Hydraulic Loading		TOC Exhaustion Time (days)	Carbon Use Rate (kg/mil L)	Organic Loading (g/kg)
		m³/hr/m²	gpm/ft²			
2.5 ft	6.2	7.3	3.0	21	92	23
5.0 ft	12.4	7.3	3.0	21	92	39
7.5 ft	18.6	7.3	3.0	28	69	60
10. ft	24.8	7.3	3.0	35	55	78

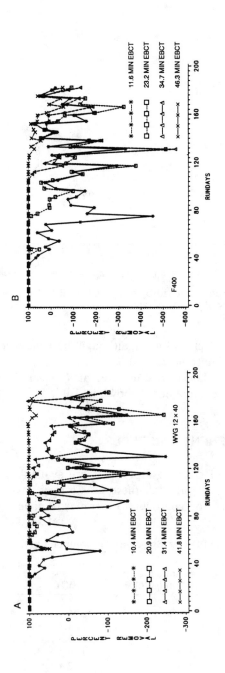

Figure 24. Percent removal of 1,2-dichloroethane—Jefferson Parish, Louisiana.

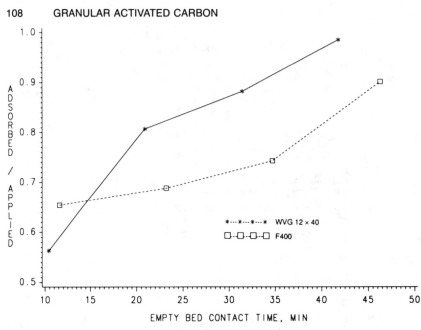

Figure 25. Ratio of 1,2-dichloroethane adsorbed/applied vs EBCT for two types of carbons—Jefferson Parish, Louisiana.

probably much greater at Jefferson Parish, although the same type of GAC was used (F400) as shown in Table 7.

Other types of specific organics were detected at Wausau. The time of GAC exhaustion for these compounds is shown in Table 8.

COMPARISON OF DIFFERENT *GAC* TYPES

GACs produced from different base products are available for use in drinking water applications.[1] Although these GACs may have the same mesh size, their performance for a particular water may be different. There-

Table 7. GAC Saturation Using Plateau (Steady-State) Criterion for 1,2-Dichloroethane

Location	Avg. GAC Influent Concentration (μg/L)	EBCT	Days to Saturation
Jefferson Parish	1.2	11.6	70
		23.2	110
		34.7	130
		46.3	160
Miami	19.9	6.2	66
		12.4	119
		18.6	171
		24.8	> 180

Table 8. GAC Exhaustion for Specific Organic Compounds Detected at Wausau, Wisconsin

Compound	Avg. GAC Influent Concentration (μg/L)	EBCT	Days to Saturation
cis-1,2-Dichloroethene	70.9	1.0	40
		3.1	60
		5.1	100
		10.4	200
		21.2	> 360
		32.2	> 360
Trichloroethene	47.9	1.0	70
		3.1	195
		5.1	> 360
		10.4	> 360
		21.2	> 360
		32.3	> 360
Tetrachloroethene	37.6	1.0	285
		3.1	360
		5.1	> 360
		10.4	> 360
		21.2	> 360
		32.3	> 360
Toluene	19.3	1.0	> 360
		3.1	> 360
		5.1	> 360
		10.4	> 360
		21.2	> 360
		32.3	> 360

fore, before a GAC is selected for use in full-scale treatment, its performance should be considered along with the cost.

During three field-scale research projects, various GACs were evaluated. At each location, the GACs were evaluated in parallel, receiving the same influent water. At Passaic Valley, New Jersey, these evaluations were done throughout the duration of the project. Before starting research at Miami and Jefferson Parish, an initial screening of GAC performance was used for carbon selection. In the GAC screening at Miami, four commercially available carbons were selected for study (Table 9).[2]

For the evaluation at Miami, glass columns 2.5 cm (1 in.) in diameter and 1.5 m (5 ft) in height with an EBCT of 6.2 min were used. Each column

Table 9. Granular Activated Carbons Evaluated During Selection Study—Miami, Florida

Manufacturer	Trade Name	Mesh Size
Calgon	Filtrasorb 400	12 × 40
Westvaco	Nuchar WVG	12 × 40
ICI	Hydrodarco 1030	10 × 30
Witco	Witcarb 950	12 × 30

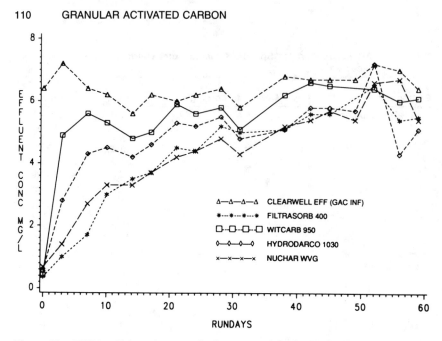

Figure 26. TOC breakthrough curves for four types of GAC—Miami, Florida.

contained 0.8 m (2.5 ft) of carbon and received clearwell finished water from the Preston Treatment Plant for the 59-day test period.

The selection criterion used at Miami was the total TOC and trihalomethane formation potential removed during the test period. Bed life (exhaustion) was determined as column breakthrough of TOC and THMFP of 3 mg/L and 200 µg/L, respectively. When using this criterion, Filtrasorb 400 had the longest bed life for TOC at 10 days. If one looks at TOC steady-state conditions, however, both Filtrasorb 400 and Nuchar WVG appear to perform equally (Figure 26).

For THMFP at 200-µg/L breakthrough, Filtrasorb 400 again had the longest bed life of 10 days and the most THMFP adsorbed within the 10-day period. Over the total 59-day test period, however, Filtrasorb 400 and Nuchar WVG again performed about the same, as shown in Figure 27 and Table 10.

Chloroform adsorption was not a GAC selection criteria at Miami. The performances of the four GACs, however, were evaluated for removal of chloroform as an indicator of low-molecular-weight-specific organic adsorption (Figure 28). The least performer for the high–molecular weight organics as indicated by TOC and THMFP was the best adsorber for chloroform (Witcarb 950).

Glass columns 7.6 cm (3 in.) in diameter and 1.8 m (6 ft) in height containing 0.8 m (30 in.) of GAC at an EBCT of 15 min were used at Jefferson Parish for evaluating the performance of various carbons. Over a

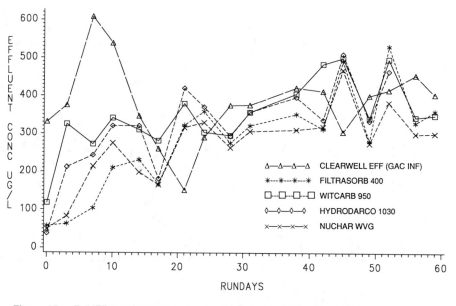

Figure 27. THMFP breakthrough curves for four types of GAC—Miami, Florida.

six-month period, five different commercially available GACs were evaluated (Table 11).[3] Sand filter effluent from Jefferson Parish's treatment plant was applied to each column.

Parameters monitored during this evaluation were nonvolatile TOC (NVTOC), instantaneous THMs, THMFP, 1,2-dichloroethane, electron capture total area (ECTOAR),* and flame ionization total area (FITOAR).* Effluent NVTOC for each GAC is presented in Figure 29. Visual observation of this figure for selection of the best performer is difficult, although WVG seems to have performed slightly better than the others. This is reaffirmed by the average NVTOC effluent concentrations:

Table 10. Trihalomethane Formation Potential Adsorbed by Four GACs—Miami, Florida

	THMFP Applied to GAC, mg		THMFP Adsorbed, mg	
Type of GAC	10 days	59 days	10 days	59 days
Filtrasorb 400	172	2002	92.6	524
Nuchar WVG	172	2002	60.3	545
Hydrodarco 1030	172	2002	23.7	314
Witcarb 950	172	2002	15.1	160

*Total area under chromatogram, excluding the solvent front, of the sample as compared to the external standards. Gas chromatographs were operated at 180°C isothermal for ECTOAR and 40–220°C at 4°C/min with a 10-min hold at 40°C for FITOAR.

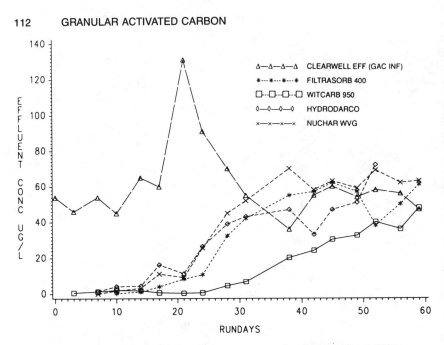

Figure 28. Chloroform breakthrough curves for four types of GAC—Miami, Florida.

WVG, 2.14 mg/L; GAC 40, 2.33 mg/L; Filtrasorb 400, 2.36 mg/L; ROW 0.8 Supra, 2.42 mg/L; and Hydrodarco 4000, 2.87 mg/L.

While evaluating the various GACs at Jefferson Parish, a higher-than-normal influent concentration of 1,2-dichloroethane was noted during the early part of the evaluation (Day 20–40). This high concentration (>10 μg/L) may have affected the length of run before the GAC became saturated. Saturation (1,2-dichloroethane) occurred for all of the GACs between Days 50 and 80. Westvaco's WVG reached saturation at Day 80 and Calgon's F400 at Day 50. The other GACs fell in between (Figure 30). There was relatively little difference in the adsorption efficiencies of the various carbons with respect to flame ionization and electron capture total area.

Prior to evaluation of the various GACs at Jefferson Parish, the initial organic load on each type of GAC was examined by Soxhlet extraction of

Table 11. Granular Activated Carbons Evaluated During Selection Study at Jefferson Parish, Louisiana

Manufacturer	Trade Name	Mesh Size
Calgon	Filtrasorb 400	12 × 40
Carborundum	GAC 40	12 × 40
ICI	Hydrodarco 4000	12 × 40
Norit	ROW 0.8 Supra®	0.8 mm diameter[a]
Westvaco	WVG	12 × 40

[a]Pelletized cylinders varying in length from 1 to 2 mm.

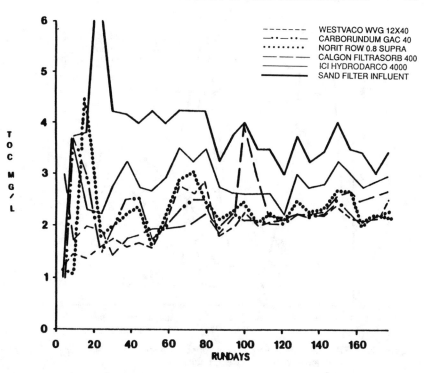

Figure 29. Effluent TOC concentrations for each type of carbon evaluated—Jefferson Parish, Louisiana.

60 g of carbon with hexane for 8 hr. The electron capture chromatograms showed essentially no contamination but a considerable amount of organics were evident on the flame ionization chromatograms. The chromatograms indicated the following GAC levels of contamination: Hydrodarco > WVG > ROW 0.8 Supra > GAC 40 > F400. The degree of initial organic contamination, however, appears to have no discernable effect on adsorption as the second most contaminated GAC, WVG, exhibited the overall best performance.

TOC removal by three carbons (HD 1030, WVG 12 × 40, and F400) evaluated simultaneously at Passaic Valley, New Jersey showed that the performance of all three was similar. Comparison of the carbons in the virgin state (Figure 31) showed that all three removed essentially the same percent of TOC through 22 days of operation. After Day 22, TOC removal with F400 was slightly better than the others.

Percentage TOC removal by once-reactivated GAC at Passaic Valley was the same for F400 and WVG 12 × 40 (Figure 32). A marked difference, however, was seen with twice-reactivated GAC (Figure 33). F400 removed TOC much better than the other carbons.

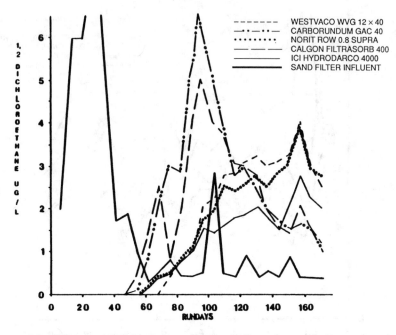

Figure 30. Effluent 1,2-dichloroethane concentrations for each type of carbon evaluated—Jefferson Parish, Louisiana.

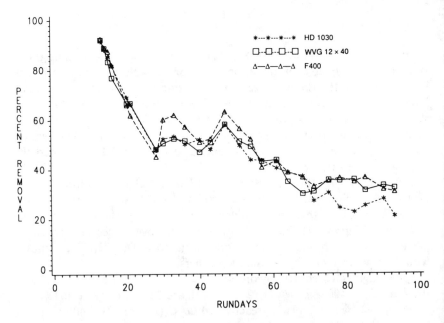

Figure 31. Percent TOC removal for virgin GAC—Passaic Valley, New Jersey.

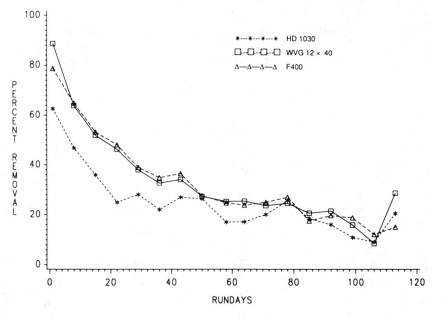

Figure 32. Percent TOC removal for once-reactivated GAC—Passaic Valley, New Jersey.

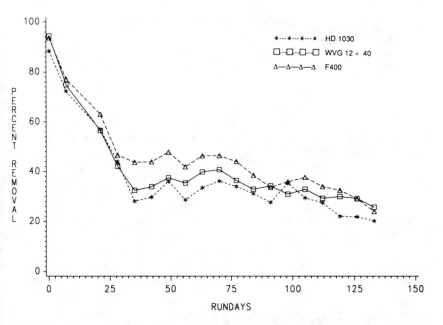

Figure 33. Percent TOC removal for twice-reactivated GAC—Passaic Valley, New Jersey.

Table 12.   Heavy Metals from Virgin Granular Activated Carbon After Acid Digestion

Metal	Filtrasorb 300[a] (mg/kg carbon)	WVG 12 × 40[b] (mg/kg carbon)
Ag	NA	BD
Al	1536	2421
As	108	3.5
Ba	50.4	54.5
Be	2.4	1.2
Bi	BD	NA
Ca	391	NA
Cd	1.2	BD
Co	BD	NA
Cr	7.2	2.7
Cu	BD	5.6
Fe	671	3361
Hg	1.9	0.2
K	NA	133
Mg	NA	144
Mn	26.4	7.8
Mo	BD	NA
Na	NA	425
Ni	12	6.3
Pb	16.8	BD
Sb	BD	NA
Se	BD	BD
Sn	BD	BD
Sr	72	BD
Ti	240	NA
Tl	BD	NA
Zn	1.2	2.4
V	24	BD

[a]25 g of GAC digested with HCl.
[b]0.1 g of GAC digested with HNO_3.
NA: not analyzed.
BD: below detection.

Table 13.   Concentration of Metals in GAC Effluents During Pilot Plant Study at Evansville, Indiana

Metal	GAC Influent (mg/L)	GAC Effluent[a] (mg/L)	GAC Effluent[b] (mg/L)
Arsenic	< 0.001	< 0.001	< 0.001
Barium	< 0.02	0.12	< 0.02
Cadmium	< 0.003	< 0.003	< 0.003
Chromium	< 0.002	< 0.002	< 0.002
Iron	0.02	0.10	0.10
Lead	0.04	< 0.02	0.08
Manganese	< 0.1	< 0.1	< 0.1
Mercury	< 0.0008	< 0.0008	< 0.0008
Selenium	< 0.0005	< 0.0005	< 0.0005
Silver	< 0.003	< 0.003	< 0.003

[a]Virgin GAC effluent. Contactor contained virgin GAC for all runs.
[b]Virgin GAC effluent. Contactor contained virgin GAC for first run.

HEAVY METALS IN *GAC*

The possibility of heavy metals leaching from GAC continues to be raised, although previous studies have shown that a problem does not exist.[4] Under stress conditions, such as acid refluxing, metals are extracted from virgin GAC. Directo, Chen, and Miele[5] refluxed 25 g of Calgon Filtrasorb 300 for 24 hr with 6N HCl and rinsed with distilled water. Concentrations of metals found are shown in Table 12. Another type of GAC (WVG 12 × 40) was digested with nitric acid, which also yielded high concentrations of heavy metals, also shown in Table 12.[3]

Under normal operating conditions during water treatment, no heavy metals were detected leaching from GAC. Effluent metal concentrations from two parallel columns containing virgin HD 10 × 30 were comparable to influent concentrations applied to the same GAC at Evansville, Indiana (Table 13).[6] Average GAC effluent concentrations for iron and manganese during 93 days of operation at Passaic Valley, New Jersey for HD 10 × 30, WVG 12 × 40, and F400 were 0.02, 0.03; 0.02, 0.02; and 0.02, 0.02 mg/L, respectively, for influent concentrations of 0.03 mg/L iron and 0.04 mg/L manganese.

REFERENCES

1. Althouse, V. E., and D. J. Treskon. *Chemical Economics Handbook on Activated Carbon* (Stanford Research Institute, August 1977).
2. Wood, P. R. et al. "Pilot-Plant Project for Removing Organic Substances from Drinking Water," Final Report, USEPA-DWRD Project CR806890 (January 1984).
3. Koffskey, W. E. "Alternative Disinfectants and Granular Activated Carbon Effects on Trace Organic Contaminants," Final Report, USEPA-DWRD Project CR806925 (April 1987).
4. Love, Jr., O. T., and J. M. Symons. "Operational Aspects of Granular Activated Carbon Adsorption Treatment," U.S. EPA (July 1978).
5. Directo, L. S., C. Chen, and R. P. Miele. "Independent Physical Chemical Treatment of Raw Sewage," U.S. EPA Report 600/2-77-137 (August 1977).
6. Lykins, Jr., B. W., M. Griese, and D. B. Mills. "Chlorine Dioxide Disinfection and Granular Activated Carbon Adsorption," U.S. EPA Report 600/2-82-051 (May 1983).

CHAPTER 5

Performance of Virgin GAC

INTRODUCTION

Selection of various types of GAC used at 9 locations was based on different criteria. The major reason for selecting a particular GAC was economics. Normally, a low bidder was chosen at each site. There were, however, occasions when a GAC type had not been evaluated and a determination was made to forego the bidding process in favor of a preselected GAC. Also, for some projects, a GAC was selected based on performance when evaluated in parallel with various other types of GAC.

Because different GACs were evaluated, they will be examined separately. Also, different contaminants were detected at various locations, making individual project evaluation necessary.

JEFFERSON PARISH, LOUISIANA

Several different operational modes were evaluated at Jefferson Parish: adsorption, filter adsorption, and pilot column simulation. The performance of virgin carbon used in the adsorption phase will be presented in this section.

The adsorber consisted of 30 in. of Westvaco WVG 12 × 40 GAC evaluated for three six-month operations (Phases I, IIA, and IIB). One other six-month study (Phase III) evaluated Calgon Filtrasorb 400, again using 30 in. of GAC. The adsorber received sand filter effluent after conventional treatment using monochloramine as the predisinfectant. Operating conditions for the four phases are shown in Table 1.

Table 1.   Operating Conditions for Adsorber—Jefferson Parish, Louisiana

Phase	Date	Average Flow Rate (gpd)[a]	Hydraulic Loading (gpm/ft²)[b]	Average EBCT (min)
I	2/7/77–8/5/77	0.41×10^6	0.75	21.9
IIA	11/4/77–4/28/78	0.40×10^6	0.73	23.7
IIB	7/10/78–1/10/79	0.59×10^6	1.07	17.0
III	4/6/79–10/5/79	0.52×10^6	0.95	18.8

[a]gpd $\times (0.0438 \times 10^{-6})$ = m³/sec.
[b]gpm/ft² $\times$ 2.4 = m/hr.
[c]Phase I, IIA, IIB—Westvaco WVG 12 $\times$ 40.
[d]Phase III—Calgon Filtrasorb 400.

Although numerous organic parameters (18 volatile and 66 nonvolatile) were monitored, only a few were detected continuously and then at microgram- or nanogram-per-liter levels. The most prevalent low-molecular-weight specific organic other than the trihalomethanes was 1,2-dichloroethane. The concentration of this compound was not reduced during conventional treatment.

GAC did, however, remove 1,2-dichloroethane until saturation, whereupon desorption occurred because of temporary decreases in the influent concentration. The occurrence of this desorption varied from 70 to 160 days of GAC filter operation (Figure 1) for applied concentrations ranging from 0.1 to 24 μg/L.

Another prevalent compound in the water at Jefferson Parish was atrazine. This chlorinated herbicide, like 1,2-dichloroethane, was not reduced in concentration during conventional treatment. Raw water concentrations ranged from < 0.1 to 5360 ng/L with variability related to water temperature and seasonal effects. Higher concentrations were noted at higher temperatures (seasonal trend). The GAC adsorber effectively removed atrazine to below detection limits in most cases and no desorption occurred up to 180 days of operation, as shown in Figure 2.

Other chlorinated hydrocarbon insecticides (CHI) monitored occurred at average concentrations below 5 ng/L. Therefore, they were combined to evaluate the cumulative GAC adsorption effect. The insecticides included in this evaluation are shown in Table 2, along with their frequency of occurrence for the sand filter effluent (GAC influent). Although the summed CHI concentrations were lower than atrazine concentrations, GAC performance was similar in that the CHI was essentially removed without noticeable desorption after 180 days of operation (Figure 3).

TOC, a general parameter found in all surface water systems, was also effectively reduced by GAC. For up to about 100 days of operation, steady-state conditions occurred for WVG 12 $\times$ 40 at an influent concentration of 3-4 mg/L TOC. However, the Filtrasorb 400 (Phase III), at an influent TOC concentration of about 3 mg/L, showed a continual rise in the adsor-

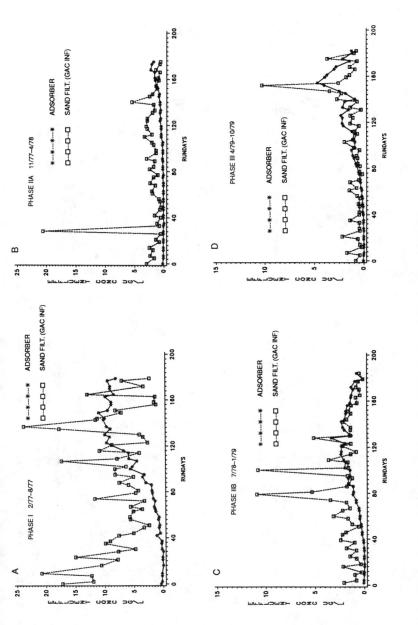

Figure 1. Concentration of 1,2-dichloroethane for GAC adsorber, Jefferson Parish, Louisiana.

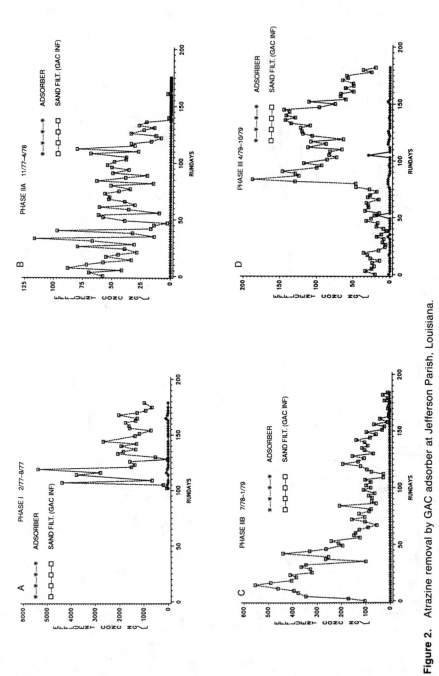

Figure 2. Atrazine removal by GAC adsorber at Jefferson Parish, Louisiana.

Table 2. Frequency of Occurrence and Concentration Ranges in the Sand Filter Effluent for all Chlorinated Hydrocarbon Insecticides Monitored During Each Phase—Jefferson Parish, Louisiana

Chlorinated Hydrocarbon Insecticides	Phase I		Phase IIA	
	Frequency of Occurrence (%)	Concentration Range (ng/L)	Frequency of Occurrence (%)	Concentration Range (ng/L)
Alpha-BHC	100	0.2–8.4	100	0.9–52.0
Beta-BHC	30	0–10.3	0	0–0
Gamma-BHC	98	0–7.3	100	0.3–5.4
Hexachlorobenzene	97	0–2.0	75	0–2.3
Heptachlor	41	0–2.0	21	0–0.5
Alpha-chlordene	19	0–1.1	34	0–0.2
Beta-chlordene	42	0–1.1	28	0–0.5
Gamma-chlordene	58	0–2.1	42	0–1.9
Aldrin	68	0–0.9	55	0–1.0
Heptachlor epoxide	91	0–1.5	100	0.2–1.2
Alpha-chlordane	97	0–1.2	100	0.1–0.7
Gamma-chlordane	98	0–2.3	100	0.2–0.8
Ortho,para'-DDE	8	0–0.4	19	0–0.3
Para,para'-DDE	32	0–0.2	82	0–0.2
Endosulfan-1	16	0–0.2	15	0–0.1
Endosulfan-2	58	0–4.0	35	0–1.6
Dieldrin	100	0.1–11.6	100	0.4–3.4
Ortho,para'-DDD	13	0–1.3	69	0–0.6
Para,para'-DDD	32	0–1.7	51	0–0.5
Endrin	100	0.04–8.2	100	0.2–2.9
Ortho,para'-DDT	6	0–0.9	31	0–0.3
Para,para'-DDT	39	0–55.5	22	0–2.2
Photodieldrin	61	0–2.9	47	0–1.3
Methoxychlor	47	0–10.7	17	0–16.4
Mirex	36	0–10.4	44	0–2.7
Pentachloronitrobenzene	—	—	—	—
DCPA (Dimethyl-2,3,5,6-tetrachloro-terephthalate)	—	—	—	—
Oxy-chlordane	—	—	—	—
Meta,para'-DDD	—	—	—	—
Perthane	—	—	—	—

Table 2. (cont'd.)

Chlorinated Hydrocarbon Insecticides	Phase IIB		Phase III	
	Frequency of Occurrence (%)	Concentration Range (ng/L)	Frequency of Occurrence (%)	Concentration Range (ng/L)
Alpha-BHC	100	1.0–12.0	100	0.8–4.4
Beta-BHC	1	0–4.8	69	0–12.4
Gamma-BHC	100	0.4–9.1	100	0.3–2.0
Hexachlorobenzene	7	0–8.1	87	0–17.2
Heptachlor	96	0–1.1	68	0–0.3
Alpha-chlordene	70	0–0.5	2	0–0.2
Beta-chlordene	63	0–0.2	27	0–0.7
Gamma-chlordene	23	0–0.2	0	0–0
Aldrin	10	0–2.3	39	0–6.1
Heptachlor epoxide	100	0.2–3.1	98	0–1.9
Alpha-chlordane	100	0.3–2.4	100	0.2–1.0
Gamma-chlordane	100	0.2–1.6	100	0.1–1.0
Ortho,para'-DDE	76	0–0.5	30	0–1.0
Para,para'-DDE	75	0–0.3	82	0–0.1
Endosulfan-1	0	0–0	0	0–0
Endosulfan-2	0	0–0	64	0–2.2
Dieldrin	100	0.9–7.4	100	1.5–7.1
Ortho,para'-DDD	87	0–2.6	59	0–1.9
Para,para'-DDD	55	0–0.4	78	0–0.5
Endrin	100	0.1–8.4	98	0–14.4
Ortho,para'-DDT	3	0–0.4	8	0–0.3
Para,para'-DDT	5	0–1.9	22	0–0.2
Photodieldrin	22	0–1.8	16	0–10.7
Methoxychlor	0	0–0	16	0–5.8
Mirex	0	0–0	11	0–6.1
Pentachloronitrobenzene	77	0–0.1	53	0–0.5
DCPA (Dimethyl-2,3,5,6-tetrachloro-terephthalate)	100	0.1–3.7	98	0–5.8
Oxy-chlordane	0	0–0	5	0–0.4
Meta,para'-DDD	0	0–0	12	0–0.9
Perthane	55	0–38.5	0	0–0

ber effluent levels for 160 days with some indication of steady state after that time of operation (Figures 4 and 5).

The phthalates, n-alkanes, and substituted benzene derivatives at the nanogram-per-liter level did not appear to be adsorbed by GAC. Some of these substances did, however, exhibit a low degree of constant removal suggesting natural biological decomposition. The compounds evaluated are presented in Table 3.

CINCINNATI, OHIO

Many specific organics were investigated at the Cincinnati site, but most were present in low concentrations near the detection limits. For example, Table 4 shows the purgeable halogenated and nonhalogenated organics that were detected in the GAC effluent of two contactors. The inability to draw conclusions on GAC effectiveness for removing various specific organics was also compounded by their infrequent occurrence. This was illustrated by tentatively identified compounds (GC/MS) for the acid extracts from one of the GAC contactors (Contactor D).

Table 5 shows that of the 45 acid extract compounds identified, only seven had a percentage of occurrence that was 30% or greater. Of these seven, dibromochloromethane, 1,3,5-trimethyl-1,3,5-triazine-2,4,6-(1H, 2H,3H)-trione, nitrobenzene, and tributyl ester phosphoric acid appeared to be well adsorbed, while 3,3,3-trichloro-1-propene, 2-cyclohexene-1-1, and 7-oxabicyclo(4.0.1)heptane removals varied from marginal to not adsorbed.

Grob closed loop stripping analysis (CLSA) was performed on one GAC contactor (A) influent and effluent samples at startup and at approximately four-week intervals for a total of 11 samples.[1] Approximately 225 compounds were identified, with a daily average of 106 and a range of 84 to 130. Except for the trihalomethanes, all concentrations were in the low ng/L range. Tables 6 and 7 present a data summary of the percent removals by GAC for 50 organic compounds. These compounds were identified most frequently (on 50% or more of the sample days) and occurred at concentrations greater than 10 ng/L on at least one of those days.

For the most part, GAC removed the identified substances. The degree of removal and length of effectiveness, however, varied from compound to compound. Some compounds, such as tetrachloroethene, 1,2-dichlorobenzene, and hexachloroethane, were well adsorbed (85–100%) by GAC over the entire sample period. Other compounds (e.g., diisopropyl ether, benzene, and carbon tetrachloride) were initially adsorbed (85–100%) but desorbed with time. A few compounds (toluene, ethylbenzene, 1,2,4-trimethylbenzene, and 1,2,3,4-tetramethylbenzene) were adsorbed only to a limited extent by GAC. Concentrations of two compounds (nonanal and

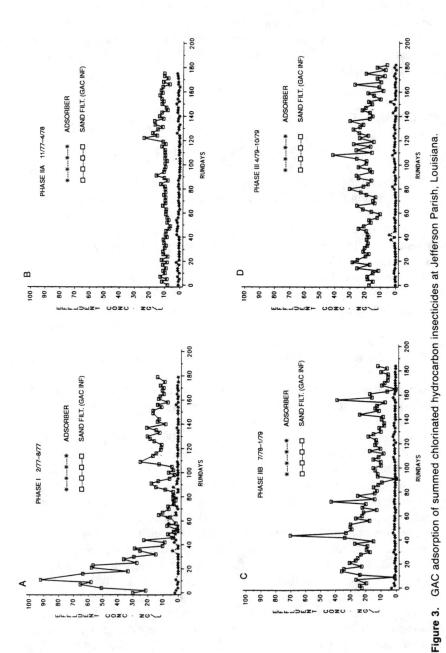

Figure 3. GAC adsorption of summed chlorinated hydrocarbon insecticides at Jefferson Parish, Louisiana.

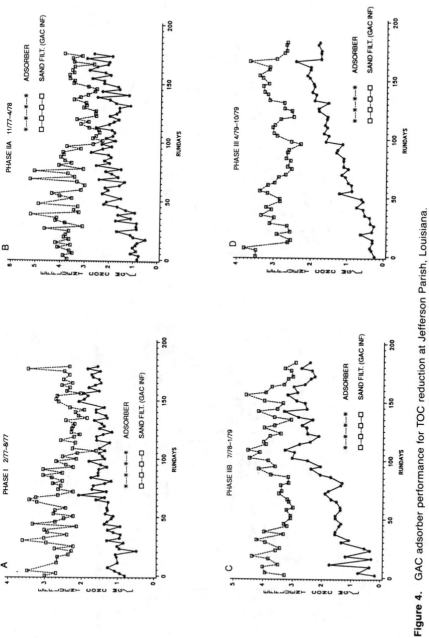

Figure 4. GAC adsorber performance for TOC reduction at Jefferson Parish, Louisiana.

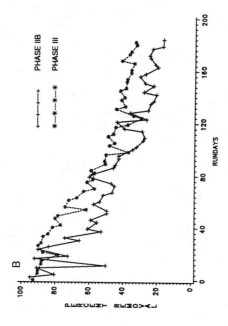

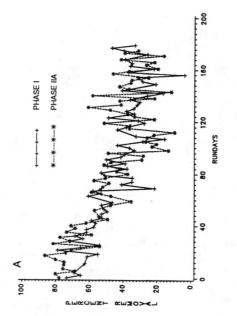

Figure 5.   Percent TOC removal at Jefferson Parish, Louisiana.

Table 3. Phthalates, N-Alkanes, and Substituted Benzene Derivatives Monitored During Each Phase at Jefferson Parish, Louisiana

Compound	Phase I		Phase IIA	
	Frequency of Occurrence (%)	Concentration Range (ng/L)	Frequency of Occurrence (%)	Concentration Range (ng/L)
Phthalates				
bis[2-Ethylhexyl] phthalate	64	0–16.5	34	0–435.
Dimethylphthalate	31	0–39.4	18	0–15.2
Diethylphthalate	39	0–40.3	94	40.0–89.4
Dipropylphthalate	6	0–4.2	1	0–0.4
Diisobutylphthalate	28	0–12.9	32	0–25.2
Dibutylphthalate	39	0–172.	93	0–120.
N-Alkanes				
N-Decane	6	0–13.7	42	0–216.
N-Undecane	24	0–15.3	34	0–9.8
N-Dodecane	34	0–15.3	57	0–6.6
N-Tridecane	12	0–2.8	44	0–32.7
N-Tetradecane	8	0–7.6	36	0–1.6
N-Pentadecane	12	0–2.3	42	0–1.6
N-Hexadecane		NA[a]		NA
N-Heptadecane		NA		NA
N-Octadecane		NA		NA
N-Nonadecane		NA		NA
N-Eicosane		NA		NA
Substituted Benzene Derivatives				
Benzonitrile		NA		NA
Benzyl cyanide		NA		NA
2-Nitrotoluene		NA		NA
Ethylbenzoate		NA		NA
Ortho-chloronitrobenzene		NA		NA
Nitrobenzene	38	0–276.	72	0–82.2
Isophorone	23	0–19.0	55	0–10.9

aNA: Not analyzed.

Table 3. (cont'd.)

Compound	Phase IIB		Phase III	
	Frequency of Occurrence (%)	Concentration Range (ng/L)	Frequency of Occurrence (%)	Concentration Range (ng/L)
Phthalates				
bis[2-Ethylhexyl] phthalate	96	0–48.0	100	1.1–170.0
Dimethylphthalate	3	0–2.6	30	0–8.6
Diethylphthalate	100	6.1–346.0	100	4.2–47.0
Dipropylphthalate	2	0–5.1	1	0–0.6
Diisobutylphthalate	77	0–33.3	100	0.8–16.0
Dibutylphthalate	98	0–211.0	100	8.0–279.0
N-Alkanes				
N-Decane	1	0–4.7	72	0–24.8
N-Undecane	0	0–0	81	0–6.7
N-Dodecane	42	0–116.0	89	0–77.6
N-Tridecane	2	0–2.0	39	0–20.0
N-Tetradecane	76	0–29.3	26	0–17.6
N-Pentadecane	48	0–5.4	46	0–20.9
N-Hexadecane	62	0–23.4	79	0–7.4
N-Heptadecane	95	0–18.5	100	1.0–14.1
N-Octadecane	61	0–17.7	84	0–8.3
N-Nonadecane	32	0–3.6	74	0–8.9
N-Eicosane	25	0–11.4	97	0–12.4
Substituted Benzene Derivatives				
Benzonitrile	2	0–9.7	86	0–43.4
Benzyl cyanide	5	0–4.7	68	0–18.6
2-Nitrotoluene	30	0–11.5	93	0–29.8
Ethylbenzoate	1	0–0.5	25	0–9.9
Ortho-chloronitrobenzene	95	0–230.0	100	1.0–193.0
Nitrobenzene	95	0–336.0	88	0–51.3
Isophorone	68	0–41.9	73	0–148.0

Table 4. Purgeable Organics Detected in GAC Influent or Effluent at Cincinnati, Ohio

	Contactor C[a]						Contactor D[a]					
	Influent (µg/L)			Effluent (µg/L)			Influent (µg/L)			Effluent (µg/L)		
Compound	min.	max.	avg.	min.	max.	avg.	min.	max.	avg.	min.	max.	avg.
Halogenated												
Carbon tetrachloride	N[b]	N	N	N	N	N	N	0.2	N	N	N	N
1,1,1-Trichloroethane	N	N	N	N	N	N	N	0.2	N	N	0.2	N
Dichloromethane	N	N	N	N	0.2	N	N	0.6	0.2	N	0.5	0.2
Tetrachloroethene	N	0.2	N	N	N	N	N	0.2	0.1	N	N	N
Trichloroethene	N	0.2	0.1	N	N	N	N	0.2	0.1	N	N	N
Chlorobenzene	N	N	N	N	N	N	N	0.2	N	N	N	N
o-Dichlorobenzene	N	N	N	N	N	N	N	0.2	N	N	0.2	N
Nonhalogenated												
Benzene	0.2	1.0	0.6	0.2	0.6	0.4	N	1.2	0.3	0.2	0.8	0.3
o-Xylene	N	0.2	N	N	0.2	N	N	N	N	N	N	N
Ethylbenzene	N	0.2	N	N	0.2	0.1	N	13.4	1.7	0.2	0.8	0.1
Toluene	N	0.9	0.2	N	0.2	N	N	1.2	0.4	N	2.8	0.4
Hexane	N	0.6	0.2	N	0.7	0.1	N	1.5	0.2	N	1.0	0.1
Tetralin	N	N	N	N	N	N	N	2.4	0.1	N	0.9	N

[a]Each contactor had a hydraulic loading of 7.4 gpm/ft² and an EBCT of 15.2 min, and used WVG 12 × 40 GAC.
[b]N: Not detected (below detection limit of 0.1 µg/L).

Table 5. Acid Extract Compounds Tentatively Identified by GC/MS From Contactor D—Cincinnati, Ohio

Compound	Percent of Occurrence			
	Inf[a]	D7[b]	DE[c]	Blank[d]
Dibromochloromethane	70	11	N[e]	N
3,3,3-Trichloro-1-propene	50	22	20	N
1,3,5-Trimethyl-1,3,5-triazine-2,4,6- (1H,2H,3H)-trione	40	N	N	N
7-Oxabicyclo(4.0.1)heptane	30	33	40	10
2-Cyclohexene-1-one	30	22	10	N
Nitrobenzene	30	N	N	N
Tributyl ester phosphoric acid	30	N	N	N
Diethyl ester-1,2-benzenedicarboxylic acid	20	11	20	10
2-Methyl-1,4-dinitrobenzene	20	N	N	N
Tribromomethane	20	N	N	N
Benzoic acid	10	N	N	N
Bromodichloromethane	10	N	N	N
1-Chloro-2-butene	10	N	N	N
Butane	10	N	N	N
Cyclohexene	10	11	N	10
Decyl ester nitric acid	10	N	N	N
1,4-Dichlorobutane	10	11	10	N
1,1-Dichloropropane	10	N	N	N
1,1-Dichloro-2-propanone	10	N	N	N
Dimethyl-1,4-dioxalane	10	N	N	N
1,2-Dimethylpentane	10	N	N	N
Ethylbenzene	10	N	10	N
3-Ethyl-4-methylfurandiene	10	N	N	N
3-Methyl-2-butadiene	10	N	N	10
5-Methylnonane	10	N	N	N
Nitrocyclopentane	10	N	N	N
2-Nitropentane	10	N	N	N
1-Pentyne	10	N	N	N
1-H-Pyrrole	10	11	N	N
1-H-1,2,4-Triazolediamine	10	N	10	N
2,2,3-Trimethylbutane	10	11	N	N
2,2,3-Trimethylpentane	10	N	N	N
Cyclobutanal	N	11	N	N
2-Cyclohexene-1-ol	N	11	N	10
Cyclopropane	N	11	N	N
2,3-Dimethyl-1-butene	N	11	N	N
3,3-Dimethylhexane	N	11	N	N
4-Ethylheptane	N	11	N	N
2-Methylnaphthalene	N	11	10	N
2,4,8-Trimethylnonane	N	11	N	N
1,3-Dimethylbenzene	N	N	10	N
2-Hexene-3-one	N	N	10	N
2,2,5-Trimethylhexane	N	N	10	N
Dibutyl ester ethanoic acid	N	N	N	10
2,2-Proanylchloride	N	N	N	10

[a]Influent to contactor D—occurrence based on 10 samples.
[b]Sample collected after 7 ft of GAC contact—occurrence based on 9 samples.
[c]Contactor D effluent (15 ft of GAC)—occurrence based on 10 samples.
[d]Redistilled solvent with internal standard added.
[e]N: not detected.

Table 6. Percent Removal Data Summary of Grob CLSA Results for Contactor A, Cincinnati, Ohio

Compound	Runday				
	1	34	64	113	134
Diisopropylether	a	100	100	100	d
1,1,1-Trichloroethane	e	95	87	94	98
Benzene	a	100	a	100	99
Carbon tetrachloride	100	94	93	100	100
Cyclohexane	g	21	f	100	c
1,2-Dichloropropane	a	100	f	f	f
Trichloroethene	a	98	98	100	100
Methylcyclohexane	d	a	a	100	f
4-Methyl-2-pentanone	a	f	100	100	f
Toluene	38	72	-3	59	100
Hexanal	-100	9	34	e	18
Tetrachloroethene	100	100	98	100	100
Chlorobenzene	a	100	92	100	100
Ethylbenzene	25	c	22	62	99
1,3- and 1,4-Dimethylbenzene	38	51	2	20	98
Styrene	100	100	40	100	100
1,2-Dimethylbenzene	40	58	17	59	99
Isopropylbenzene	37	50	33	67	99
Propylbenzene	33	36	0	62	98
Ethyl-4-methylbenzene	43	48	17	32	98
1,3,5-Trimethylbenzene	33	c	e	e	e
1,2,4-Trimethylbenzene	27	43	11	46	99
Octanal	b	d	-77	39	d
1,4-Dichlorobenzene	100	100	100	100	d
1,2,3-Trimethylbenzene	33	46	100	100	99
1,2-Dichlorobenzene	100	100	99	100	98
1,3-Diethylbenzene	40	d	e	67	99
1,4-Diethylbenzene	d	-100	g	c	e
5-Ethyl-1,3-dimethylbenzene	60	e	0	57	99
Hexachloroethane	100	99	98	100	99
2-Ethyl-4,4-dimethylbenzene	80	50	d	b	e
4-Ethyl-1,2-dimethylbenzene	50	e	20	50	99
Nonanal	-350	-350	-150	-9	-1300
3-Ethyl-1,2-dimethylbenzene	70	e	0	b	99
1,2,3,5-Tetramethylbenzene	-50	55	e	g	99
1,3-Diethyl-5-methylbenzene	50	0	0	100	100
1,2,3,4-Tetramethylbenzene	60	50	50	83	99
Decanal	-150	-680	-120	-17	d
Dodecanal	a	b	g	100	d
2,6-bis(1,1-Dimethylethyl)2,5-cyclohexadiene-1,4-dione	67	e	70	100	e
1,1,3-Trimethyl-3-phenylindan	33	c	-10	g	d
Heptadecane	33	d	-180	e	d
Dibutylphthalate	a	c	c	c	g
Dioctylphthalate	a	a	f	100	g
Hexane	a	a	a	a	a
Methylcyclopentane	a	a	a	a	a

aNot detected.
bNot detected in influent.
cNot quantified.
dInfluent not quantified.
eEffluent not quantified.
fInfluent not quantified and effluent not detected.
gInfluent not detected and effluent not quantified.

Table 7. Percent Removal Data Summary of Grob CLSA Results for Contactor A, Rundays 162 Through 302

Compound	Runday					
	162	190	218	246	274	302
Diisopropylether	67	-114	f	g	b	g
1,1,1-Trichloroethane	56	70	f	c	-35	d
Benzene	100	100	a	40	-4	-46
Carbon tetrachloride	100	a	100	e	e	-96
Cyclohexane	90	-46	d	c	c	c
1,2-Dichloropropane	a	100	a	a	h	h
Trichloroethene	100	100	a	100	a	17
Methylcyclohexane	100	d	g	c	g	g
4-Methyl-2-pentanone	88	-166	b	a	g	a
Toluene	90	-83	42	15	-21	-75
Hexanal	63	-15	e	a	0	g
Tetrachloroethane	100	100	100	e	86	92
Chlorobenzene	100	a	a	h	c	g
Ethylbenzene	e	-286	e	c	-3900	65
1,3- and 1,4-Dimethylbenzene	71	-220	33	a	a	a
Styrene	95	h	a	a	e	e
1,2-Dimethylbenzene	84	-75	e	e	-2000	23
Isopropylbenzene	96	0	b	g	d	c
Propylbenzene	83	-33	a	g	c	c
Ethyl-4-methylbenzene	e	-25	56	100	-43	-350
1,3,5-Trimethylbenzene	c	d	c	e	e	16
1,2,4-Trimethylbenzene	68	0	36	57	-100	-24
Octanal	68	-60	e	h	44	d
1,4-Dichlorobenzene	99	100	100	e	89	86
1,2,3-Trimethylbenzene	80	0	d	e	e	-12
1,2-Dichlorobenzene	99	98	100	100	100	100
1,3-Diethylbenzene	g	h	c	g	c	c
1,4-Diethylbenzene	83	h	c	g	g	c
5-Ethyl-1,3-dimethylbenzene	e	0	d	100	e	e
Hexachloroethane	99	100	100	a	100	100
2-Ethyl-1,4-dimethylbenzene	c	g	d	e	c	e
4-Ethyl-1,2-dimethylbenzene	73	0	a	100	e	-220
Nonanal	54	-80	-23	-48	23	-15
3-Ethyl-1,2-dimethylbenzene	b	b	b	g	c	g
1,2,3,5-Tetramethylbenzene	e	0	d	e	e	0
1,3-Diethyl-5-methylbenzene	a	a	a	a	g	g
1,2,3,4-Tetramethylbenzene	87	a	b	g	e	e
Decanal	65	-76	9	-22	19	-10
Dodecanal	68	-4	-5	h	6	100
2,3-bis(1,1-Dimethylethyl)2,5-cyclohexadiene-1,4-dione	94	100	100	c	100	-52
Pentadecane	h	b	-45	-44	42	75
Diethylphthalate	29	48	g	42	e	35

Table 7.   (cont'd.)

Compound	Runday					
	162	190	218	246	274	302
2,2,4-Trimethylpenta-1,3-diolidisobutyrate	65	100	a	a	91	-2
2,5-bis(1,1-Dimethylpropyl)2,5-cyclohexadiene-1,4-dione	76	a	c	-81	c	83
1,1,3-Trimethyl-3-phenylindan	e	61	-210	21	84	g
Heptadecane	26	12	-73	57	21	14
Dibutylphthalate	c	c	a	h	80	c
Dioctylphthalate	h	a	a	a	a	a
Hexane	a	-54	f	a	-12	100
Methylcyclopentane	g	-6	f	e	41	c

aNot detected.
bNot detected in influent.
cNot quantified.
dInfluent not quantified.
eEffluent not quantified.
fNot scanned.
gInfluent not quantified and effluent not detected.
hInfluent not detected and effluent not quantified.

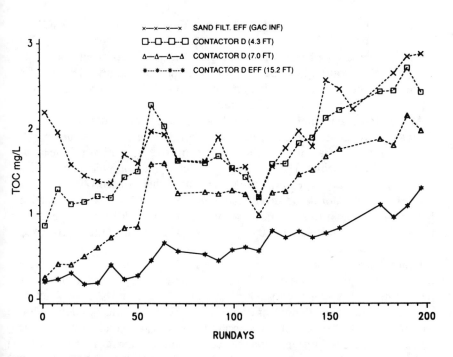

Figure 6.   TOC breakthrough curves for Contactor D, Cincinnati, Ohio.

decanal) were higher in the effluent than in the influent. It is suspected that this was caused either by the compounds leaching from the contactor liner, a result of laboratory contamination, or by bacterial degradation of adsorbed organics.

Operation of GAC Contactor A was allowed to continue for about a year to evaluate its adsorptive capacity for the various compounds listed in Tables 6 and 7. After a certain period of use, the GAC was exhausted for some of the compounds. This effect is especially prevalent in Table 7 where the minus percent removal indicates desorption occurred.

The use of TOC as a GAC performance parameter was chosen because of the variability of occurrence of specific organics at low concentrations (nanogram-per-liter level) and the impossibility of daily monitoring for all organics present in the water. Generally, TOC analysis of the GAC contactor sampling points provided a typical breakthrough curve with increased carbon depth providing lower TOC concentrations for a longer time period. Figure 6 shows a typical example where a TOC influent concentration of 2.0–2.5 mg/L was reduced to below 1.0 mg/L up to about 150 days of operation at an EBCT of 15.2 min.

HUNTINGTON, WEST VIRGINIA

In Huntington, West Virginia and Beaver Falls, Pennsylvania, GAC was an integral part of the treatment scheme for taste and odor control. The adsorptive capacity of virgin GAC in the filtration/adsorption mode was investigated at both utilities. At the time of these studies only a few specific organics, other than the trihalomethanes and their precursors, were detected in the microgram-per-liter range.

Most of the purgeable and base-neutral extractable halocarbons were detected infrequently; when detected, they were typically at or below 0.2 μg/L. Therefore, evaluation of the GAC adsorptive capacity for these compounds was difficult. There were, however, some organic compounds present sporadically in Huntington's water at higher concentrations.

Carbon tetrachloride occurred frequently. Virgin GAC was an effective barrier when high influent concentrations occurred at Week 10 of operation (13 μg/L in GAC influent and 0.4 μg/L in GAC effluent). Huntington used WVW 14 × 40 carbon.

Chlorobenzene concentrations in the microgram-per-liter range were detected infrequently in the GAC influent. When detected, concentrations ranged up to 1.0 μg/L (6th week of operation). During the 35th week of operation, the influent concentration was 0.5 μg/L and was not detected in the effluent. In 10 out of 15 sampling times, 1,4-dichlorobenzene was detected in the GAC influent at a concentration that ranged from 0.1 to 1.4 μg/L. At Week 22 of operation, the highest GAC influent concentration of

1.4 μg/L was effectively removed, with none detected in the GAC effluent. Occasionally, base-neutral extractable halocarbons were detected. Concentrations in GAC influent waters were low (< 1.0 μg/L) with no concentrations detected in the GAC effluent.

BEAVER FALLS, PENNSYLVANIA

Occasionally, carbon tetrachloride was detected in the treated water. Its presence was suspected to be a result of a contaminated chlorine supply. When detected, concentrations were typically below 0.2 μg/L, but not enough data were available to evaluate the effectiveness of GAC adsorption. Also found occasionally in the GAC influent (plant settled water) was 1,4-dichlorobenzene at concentrations ranging from 0.1 to 0.3 μg/L. Again, not enough data were available to evaluate GAC adsorption efficiency.

Unidentified base-neutral extractable halocarbons were detected periodically in the finished water but rarely in the raw water. When detected, their concentrations ranged from 0.1 to 1.2 μg/L. Some adsorption was noted during the first four months of operation, but after that time the concentration of the GAC effluents was similar to the influent.

At Beaver Falls, three virgin GAC beds were evaluated in parallel with sand filtration. The three beds were composed of F400, Filtersorb C, and HD 8 × 16.

EVANSVILLE, INDIANA

During this project, only the general parameter (TOC) was detected continuously with most of the other specific organics detected sporadically at the detection limit of the instrumentation used.

The applied TOC concentration to the GAC contactors was between 1.6 and 3.0 mg/L, with the lowest concentrations seen during the months of March through May. The virgin GAC effluent, as evaluated by the percentage TOC removed, showed that at the beginning of each run this percent was about 80 for two out of three runs, and by Day 63, steady-state conditions had occurred for all runs (Figure 7).

Occasionally, tetrachloroethylene was detected both in the raw and finished water, with concentrations decreasing during treatment. Average instantaneous raw water concentrations ranged from 0.1 to 0.4 μg/L. GAC effluent concentrations were typically less than 0.1 μg/L.

MANCHESTER, NEW HAMPSHIRE

Because of the pristine nature of the source water (Lake Massabesic) no specific organics other than the THM constituents were detected at Man-

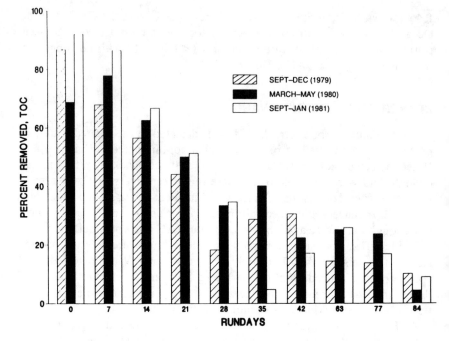

Figure 7. Percent TOC removed through virgin GAC, Evansville, Indiana.

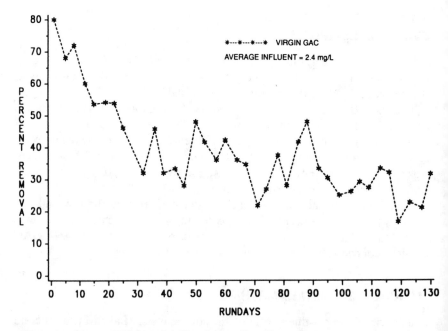

Figure 8. Percent TOC removal, Manchester, New Hampshire.

Table 8.   Average Concentration of Specific Halogenated Organics in Raw Water, Miami, Florida

Chemical Name	Average Concentration in μg/L Raw Water		
	ED1R[a]	ED3[b]	ED4[c]
Vinyl chloride	0.8	6.9	12.8
Methylene chloride	0.1	0.1	0.5
trans-1,2-Dichloroethene	1.3	2.0	1.5
1,1-Dichloroethane	0.3	0.6	0.6
cis-1,2-Dichloroethene	29.0	26.9	25.6
1,1,1-Trichloroethane			
1,2-Dichloroethane	0.1[d]	0.2[d]	0.1[d]
Carbon tetrachloride			
Trichloroethylene	0.1	0.4	0.3
Tetrachloroethylene	0.1	N[e]	N
Chlorobenzene	0.2	1.3	1.1
p-Chlorotoluene	0.1	0.2	N
m-Dichlorobenzene			
p-Dichlorobenzene	1.1[f]	1.0[f]	0.7[f]
o-Dichlorobenzene			

[a]ED1R: repeat of study 1 (Jan. 18–May 20, 1977).
[b]ED3: study 3 (Aug. 26–Oct. 18, 1977).
[c]ED4: study 4 (Nov. 1, 1977–Mar. 3, 1978).
[d]Values were summed because the three peaks overlapped, making separate analysis impractical with available instrumentation.
[e]N: nil.
[f]Summed because available chromatographic techniques could not separate isomers.

chester. TOC removal by GAC adsorption proceeded according to the typical breakthrough curve. About 80% removal was accomplished at startup, and after approximately 80 days of operation steady-state removal conditions were reached at around 30% (Figure 8). Influent TOC concentrations applied to the GAC ranged from 1.8 to 2.8 mg/L, with most data points in the 2.5 mg/L area. Manchester used WVW 8 × 30 carbon.

MIAMI, FLORIDA

Several halogenated organics were detected in the Miami raw water, as shown in Table 8. In most cases the concentration of these organics was reduced during treatment (Table 9). There were some, however, that did increase in concentration, such as the summation of 1,1,1-trichloroethane, 1,2-dichloroethane, and carbon tetrachloride. The combined average concentration of these three increased more than any of the other halogenated compounds detected (77%) indicating that one or all of these compounds was a contaminant added or produced during treatment. Carbon tetrachloride is suspect. Nonetheless, after 24.8 min EBCT, no breakthrough was seen after 122 days of GAC operation.

The halogenated organic compound occurring in highest concentration was *cis*-1,2-dichloroethene, which had a raw water average concentration of

Table 9.   Average Concentration of Specific Halogenated Organics in Finished Water, Miami, Florida

	Average Concentration in μg/L		
Chemical Name	ED1Rª	ED3ᵇ	ED4ᶜ
Vinyl chloride	0.6	5.7	8.4
Methylene chloride	NDᵈ	ND	ND
trans-1,2-Dichloroethene	0.2	1.0	0.8
1,1-Dichloroethane	0.2	0.3	0.4
cis-1,2-Dichloroethene	17.2	18.3	19.9
1,1,1-Trichloroethane			
1,2-Dichloroethane	0.7ᵉ	5.3ᵉ	7.7ᵉ
Carbon tetrachloride			
Trichloroethylene	0.2	0.1	0.7
Tetrachloroethylene	N	N	N
Chlorobenzene	0.1	0.8	0.9
p-chlorotoluene	N	0.2	0.1
m-dichlorobenzene			
p-dichlorobenzene	0.6ᵍ	0.3ᵍ	0.4ᵍ
o-dichlorobenzene			

ªED1R: repeat of study 1 (Jan. 18–May 20, 1977).
ᵇED3: study 3 (Aug. 26–Oct. 18, 1977).
ᶜED4: study 4 (Nov. 1, 1977–Mar. 3, 1978).
ᵈND: not determined.
ᵉValues were summed because the three peaks overlapped, making separate analysis impractical with available instrumentation.
ᶠN: nil.
ᵍSummed because available chromatographic techniques could not separate isomers.

25.6 μg/L. The other compounds were present at much lower concentrations. All of these organics were removed by virgin GAC for varying periods of time, as shown in Table 10. The carbon used in Miami was F400.

PASSAIC VALLEY, LITTLE FALLS, NEW JERSEY

At a short EBCT of 8 min, effective performance of the GACs used at Passaic Valley was limited. Both TOC and TOX reached a removal plateau after approximately 9 weeks of operation at 30% and 40%, respectively. Although a total of 42 organic parameters were analyzed, including chlorinated pesticides and volatile halogenated solvents, only a few were detected in the influent applied to the GAC with any consistency, and these were at the microgram-per-liter level.

These included carbon tetrachloride and tetrachloroethylene. Out of 32 observations that included 102 days of operation, carbon tetrachloride was seen in the GAC influent eight times (during Days 46–89) at a concentration range of 0.05 to 0.31 μg/L. Removal of the carbon tetrachloride by the GACs was somewhat erratic, with the GAC effluent concentration sometimes higher than the influent (Table 11).

The other volatile organic (tetrachloroethylene) was also detected sporad-

Table 10.   Duration of Removal by GAC for Halogenated Organics Detected in Miami, Florida Water

Study	EBCT (min)	Average Influent (μg/L)	Column Breakthrough (days)	Column Saturation (days)
ED1R				
cis-1,2-Dichloroethene	6.2	29	16	73
trans-1,2-Dichloroethene	6.2	1.3	66	142
1,1-Dichloroethane	6.2	0.3	21	94
1,1,1-Trichloroethane				
1,2-Dichloroethane	6.2	0.1	21	98
Carbon tetrachloride				
Trichloroethylene	6.2	0.1	77	>122
Chlorobenzene	6.2	0.2	>122	>122
ED3				
cis-1,2-Dichloroethene	6.2	18.3	18	53
Vinyl chloride	6.2	5.7	21	>53
trans-1,2-Dichloroethene	6.2	1.0	>53	>53
1,1-Dichloroethane	6.2	0.3	16	21
1,1,1-Trichloroethane				
1,2-Dichloroethane	6.2	5.3	28	58
Carbon tetrachloride				
Trichloroethylene	6.2	0.1	a	a
Chlorobenzene	6.2	0.8	>53	>53
ED4				
cis-1,2-Dichloroethene	24.8	19.9	>122	>122
Vinyl chloride	24.8	8.4	35	87
trans-1,2-Dichloroethene	24.8	0.8	>122	>122
1,1-Dichloroethane	24.8	0.4	77	87
1,1,1-Trichloroethane				
1,2-Dichloroethane	24.8	7.7	>122	>122
Carbon tetrachloride				
Trichloroethylene	24.8	0.7	a	a
Chlorobenzene	24.8	0.9	>122	>122

[a]Detected concentrations were sporadic.

ically. Its removal was consistent among the three GACs evaluated and only on Operational Day 81 was the effluent concentration higher than the influent concentration (Table 12).

THORNTON, COLORADO

Except for the four major trihalomethanes, no other specific organics were evaluated. The surrogate parameters of TOC and TOX were used to determine the effectiveness of GAC. As described previously, various treatment modes were used prior to GAC adsorption. The GAC used in both the post–sand filtration adsorption and combined filtration/adsorption schemes was Westvaco 12 × 40. The EBCT was 20 min for each of these adsorption schemes.

Table 13 shows the average concentration of TOC for four treatment modes. As indicated in this table, various conventional treatment schemes

Table 11.   Carbon Tetrachloride GAC Influent and Effluent Concentrations, Passaic Valley, New Jersey

Operation Day Observed[a]	GAC Influent (μg/L)	GAC Effluent		
		HD-1030 (μg/L)	WVW (μg/L)	F400 (μg/L)
46	0.24	0.32	0.43	0.23
53	0.18	0.31	0.31	0.30
60	0.31	0.25	0.24	0.25
67	0.15	0.03	0.02	0.03
75	0.16	0.14	0.14	0.16
81	0.07	0.16	0.17	0.11
88	0.11	0.08	0.06	0.05
89	0.05	—	0.04	0.06
Avg.	0.16	0.18	0.18	0.15

[a]Days of detectable concentrations.

Table 12.   Tetrachloroethylene GAC Influent and Effluent Concentrations, Passaic Valley, New Jersey

Operation Day Observed[a]	GAC Influent (μg/L)	GAC Effluent		
		HD-1030 (μg/L)	WVW (μg/L)	F400 (μg/L)
46	0.66	0.06	0.12	0.11
53	0.73	0.14	0.25	0.19
67	0.58	0.11	0.07	0.08
75	1.16	0.86	0.88	0.93
81	0.66	0.88	0.97	0.64
88	0.28	0.06	0.03	0.03
89	0.05	—	0.03	0.04
Avg.	0.59	0.35	0.33	0.29

[a]Days of detectable concentrations.

did reduce the raw water TOC concentration (average raw water range 4.7–5.7 mg/L and pilot plant effluent 4.1–4.6 mg/L). The GAC further reduced the TOC concentration. The range for the filter adsorber effluent for the four modes was 0.6–1.1 mg/L; for the adsorber effluent, the range was 0.8–2.1 mg/L.

TOX precursor concentrations were reduced during conventional treatment. The degree of reduction was dependent on the predisinfectant scheme used. Also, the predisinfectant scheme did impact the instantaneous TOX concentration in the pilot plant filter effluent (Table 14). Although average concentrations can be misleading, those presented in Table 14 typify what was seen during treatment. During the longer runs (6 months), the terminal TOX adsorber effluent was higher than the filter adsorber effluent, indicating that adsorbed material on the mixed-media filter or GAC was contributing to the TOX formation potential. An example of TOX reduction through

Table 13. Average Concentrations for TOC During Thornton, Colorado Project

	TOC (mg/L)			
Mode	Raw Water	Pilot Plant Filter Effluent	Filter Adsorber Effluent	Adsorber Effluent
1 (6 months)	5.1	4.1	0.9	1.3
2 (2 months)	5.7	4.5	0.9	1.2
3 (3 months)	5.5	4.6	0.6	0.8
4 (6 months)	4.7	4.3	1.1	2.1

Table 14. Average Concentrations for TOX During Thornton, Colorado Project

	TOX, μg/L			
Mode	Raw Water[a]	Pilot Plant Filter Effluent[a]	Filter Adsorber Effluent[a]	Adsorber Effluent[a]
1 (6 months)	92/597	464/475	92/171	128/266
2 (2 months)	34/121[b]	187/395	29/61	21/30
3 (3 months)	30/452	142/404	13/32	20/22
4 (6 months)	41/315	75/164	42/78	23/151

[a]Inst/Term. Inst. = instantaneous; Term. = terminal (stored 5 days, pH 8.0, ambient temperature).

[b]Cl_2 residual not maintained during 5-day storage period.

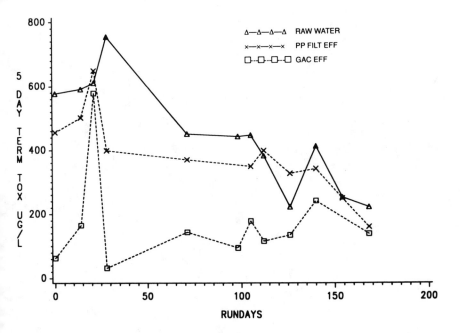

Figure 9. TOX reduction by GAC, Thornton, Colorado.

GAC is shown in Figure 9 for Mode 1. In this mode, the 5-day terminal TOX was effectively removed by GAC through 160 days of operation.

REFERENCES

1. Miller R., and D. J. Hartman. "Feasibility Study of Granular Activated Carbon Adsorption and On-Site Regeneration," EPA-600/52-82-087 (November 1982).

CHAPTER 6

Reactivation Systems

INTRODUCTION

One of the most valuable results of the field-scale experiences described previously was the use of both fluidized and infrared furnaces for reactivation of GAC onsite. Fluid bed furnaces were utilized at the Cincinnati and Manchester waterworks and an infrared furnace was utilized at the Jefferson Parish waterworks. These reactivation systems are discussed in detail in the following sections.

CINCINNATI WATER WORKS

A major objective of the Cincinnati Water Works (CWW) project was to evaluate the utilization of full-scale adsorbers and GAC filters with onsite GAC reactivation and to determine GAC losses under onsite reactivation conditons.[1]

The CWW utilizes rapid sand filtration in its treatment process. The filtration plant houses 47 filter units, each having an effective area of 130 m² (1400 ft²) with a normal operating rate of 1.7 L/sec/m² (2.5 gpm/ft² or 5 mgd).

The filters are of reinforced concrete construction, each comprised of two boxes 4.3 m (14 ft) wide by 15 m (50 ft) long, separated by a 0.76-m (2.5-ft)-wide gullet running the long dimension. The gullet serves as the filter influent and backwash discharge flume. A series of four troughs per box conducts the backwash to the gullet and serves to attain uniform effectiveness of the backwash operation. Seven filters, of most recent construction, have Leopold perforated tile bottoms and surface wash. The other 40 filters have a network of perforated cast iron pipe laterals which conduct the filtered water to collection headers beneath the filter structure.

Construction and Modification

For the purposes of this study, five of the water treatment plant's 47 filters were selected in an area (Figure 1) where the most recent filter rebuilding had taken place in order to minimize the possibility of a break-through of turbidity during the study. A single concentration of filters was considered desirable to simplify the media transport system and as a convenience in collecting samples (Figure 2).

One empty filter box was set aside for use as a storage facility for reactivated GAC in one half and virgin GAC in the other. Three GAC filters (19A, 21A, and 23A) were put into service during Phase 1 of the study using various configurations and GAC types. Another GAC filter (15A) was used during Phase 3 for direct comparison with contactor operation.

Four contactor tanks were placed in the south half of a separately built contactor building (Figure 3). Each contactor unit was basically a cylindrical shell construction of 9.55-mm (0.375-in.)–thick carbon steel plate (Figure 4).

Each tank had a "GAC carbon-in" connection at the center of the top head, and just below it a cone which distributed incoming GAC and protected the influent lateral system. At the top of each tank was an air/vacuum release which vented air as the tank was being filled and admitted air when it was being drained. A steel ladder with safety cage, extending the full height of the tank, provided access to a 50.8-cm (20-in.) manway which was located on the side of the tank above the fully charged GAC bed level. Each tank was supported by four wide flanged (WF) structural steel legs. Specifications for the contactor tanks are shown in Table 1.

Although the "A," "B," "C," and "D" labeling on the tanks and instrumentation appears to be random, it was not. This resulted from the identification being assigned in the order that each unit became available from the installing contractor and put into service.

Contactors could be charged directly from the GAC tank truck, the

Table 1. Contactor Design Specifications

Diameter	3.4 m (11 ft)
Vertical sidewall height	6.7 m (22 ft)
GAC depth	4.6 m (15 ft)
Design capacity	0.04 m³/s (1.0 mgd)
Hydraulic loading	285 L/min/m² (7.4 gpm/ft²)
Contact time	15.3 min
Design pressure	517.1 kPa (75 psig)
Test pressure	620.5 kPa (90 psig)
Backwash hydraulic loading	407 L/min/m² (10.0 gpm/ft²)

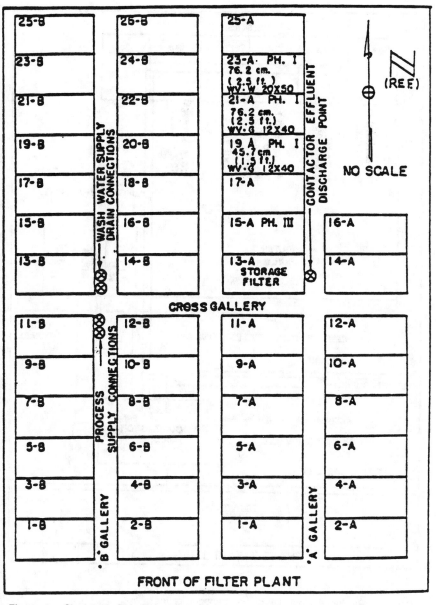

Figure 1. Cincinnati, Ohio filter gallery layout.

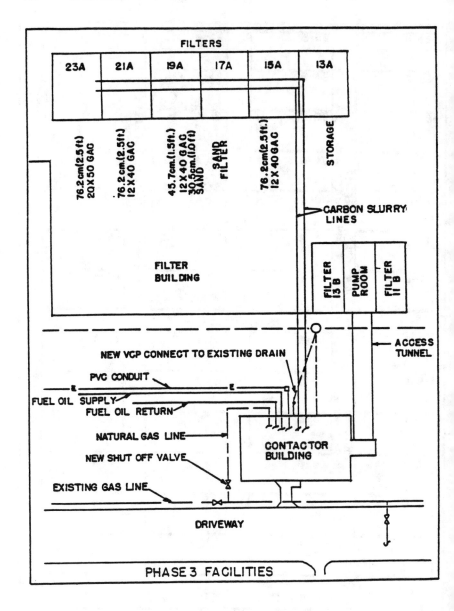

Figure 2. Cincinnati, Ohio filter carbon transport piping.

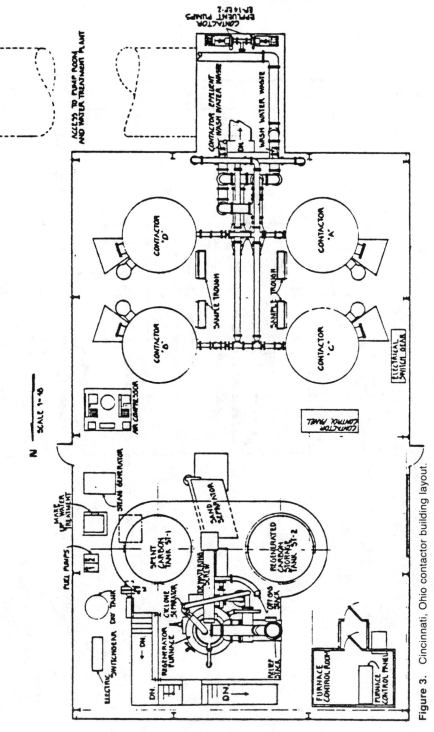

Figure 3.   Cincinnati, Ohio contactor building layout.

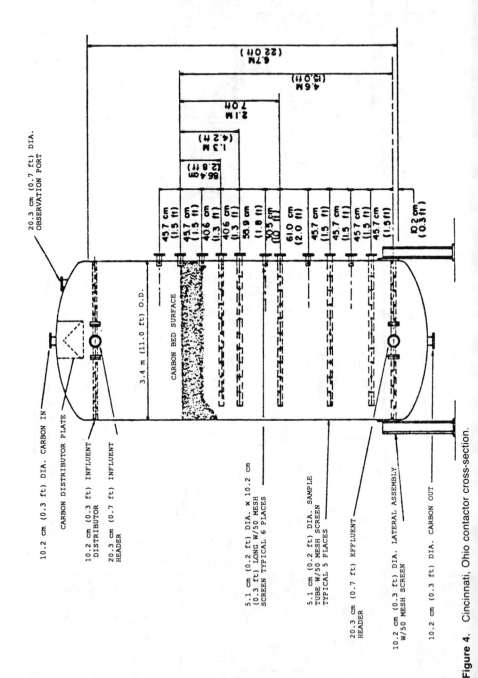

Figure 4. Cincinnati, Ohio contactor cross-section.

Table 2.   Washwater Pump Specifications

Total rated head	12.2 m (40 ft)
Capacity at rated head	82 L/sec (1300 gpm)
Minimum operating head	10.7 m (35 ft)
Maximum power at minimum operating head	16.3 kW (21.8 brake hp)
Net positive suction head (NPSH)	3.05 m (10 ft)

reactivated GAC storage tank, or the makeup GAC filter (13A). Each contactor carbon-in valve control was a pneumatic cylinder-operated valve.

Spent GAC was removed from the contactors and transferred to the spent GAC storage tank. The GAC slurry consistency was maintained at 10–15% to transfer in or out of the contactors. Water injection nozzles were welded into the GAC transport pipes to accomplish this. The carbon-out valves were pneumatic cylinder–operated valves. Control for the valves were the same as the carbon-in valves.

Backwash System

The backwash water and drain system was also interconnected with this overhead network. Valving arrangements permitted backwash water to enter the contactor through the effluent laterals and exit through the influent laterals.

The backwash pump was used to pump the backwash water through the contactor. The pump took its suction from the 91.4-cm (36-in.) washwater line that supplied the backwash water for the B-gallery sand filters. The pump was driven by an induction-type motor operating on 460 volts (three-phase). Pump specifications are listed in Table 2.

Influent Piping

The connection for supplying sand filter effluent water was previously described under modifications to Filter 11B.

From this point of connection, a 40.6-cm (16-in.) pipeline was run to the suction connection of the two 0.1-m³/sec (3-mgd) process pumps. The elevation of much of this line was above the hydraulic gradient of the flume; therefore, it became necessary to add a vacuum priming system.

One process pump served the contactors with the second acting as a standby. The process pumps were driven by a horizontal, totally enclosed, 460-volt, three-phase induction motor. The process pump specifications are shown in Table 3. The pump discharges were manifolded into a single 40.6-cm (16-in.) line going through the access tunnel to the contactor building.

Table 3. Process Pump Specifications

Total rated head	38.4 m (126 ft)
Capacity at rated head	132.5 L/s (2100 gpm)
Approximate capacity range	63.1–151.4 L/s (1000–2400 gpm)
Operating head range	29.9–39.6 m (98–130 ft)
Maximum power	64.9 kW (87 brake hp)
Pump rotation	Counterclockwise
NPSH	4.6 m (15 ft)

Effluent Piping

The water from the effluent of the contactors flowed through a 40.6-cm (16-in.) effluent line. This line was routed through the access tunnel and pump room and across the filter building, where it discharged into the A-gallery effluent flume through a connection welded to the cover plate of an access manhole near Filter 13A. The long run of pipe was planned in order to introduce the contactor effluent as far downstream as practicable to eliminate the possibility of recirculating water through the contactor that had previously been through the process.

Flowmeter.   Each contactor was equipped with a 15.2-cm (6-in.) magnetic flowmeter capable of measuring the flow of filtered water in both directions without any moving parts in the meter coming in contact with the water. The flowmeters were designed for a maximum pressure of 1172 kPa (170 psig). Each contactor had a two-pen indicating recorder with a chart and scale which read 0–0.06 m^3/sec (0–1.5 mgd). This meter also measured the rate of backwash.

Flow indicating controller.   Each contactor had a flow indicating controller to set the rate through the contactor. The flow could be controlled manually or automatically.

Pressure recorders.   Each contactor had two-pen pressure recorders and indicators with a chart and scale which read 0–700 kPa (0–100 psig). One pen recorded the pressure at the top of the contactor and the other the bottom pressure.

Instrument Air System

The instrument air system was used for the metering and control of the contactors. The compressed air equipment consisted of two heavy-duty, single-acting, two-stage, air-cooled, V-belt–driven compressors; a vertical receiver; and two refrigeration-type, self-contained air dryers. The specifications for the instrument air system are shown in Table 4. The control cabinet for the system, which contained indicating lights, switches, relays, solenoid valves, and gauges, was located immediately adjacent to the compressor.

Table 4. Instrument Air System Specifications

Compressor discharge pressure	861.9 kPa (125 psig)
Motor speed	1800 rpm
Power requirement	18.6 kW (25 hp)
Receiver test pressure	1034.2 kPa (150 psig)
Receiver volume	1.1 m³ (40 ft³)
Dryer capacity	0.84 mm (30 scfm)
Dryer power	0.38 kW (0.5 hp)
Receiver dewpoint	2 °C (35°F)

Reactivator

The fluidized bed reactivation process, sometimes referred to as "reactivation," is a vertical process where the GAC progresses downward through the reactivator counterflow to rising hot gases, which carry off volatiles as they dry the spent GAC and pyrolyze the adsorbate.

The reactivation vessel was 6.9 m (22.5 ft) tall and 2.1 m (7 ft) at its widest diameter (Figure 5). Typically, the outer steel shell was lined with 10 cm (4 in.) of mineral wool blocks covered with a cast-in-place refractory of up to 0.36 m (14 in.) in thickness. From bottom to top, it was divided into three compartments which house four functional areas (Figure 5). The bottom section was the combustion chamber, into which a stoichiometrically balanced stream of fuel and oxygen (air) flowed. These expanding gases of combustion provided the heat and fluidizing medium. Steam was also injected into this chamber; while it was part of the reactivation process, the volumes were predicated on the need for additional fluidizing gases. Temperatures reached approximately 1040°C (1900°F) in the combustion chamber, which had a water-sealed pressure relief vent. The burners were of a dual fuel type which used either fuel oil or natural gas.

The reactivation zone was separated from the combustion chamber by a stainless steel diaphragm fitted with a number of nozzles to distribute the gases uniformly over the cross-section of the furnace. It also supported the GAC when at rest. A series of weir plates rested on the diaphragm plate to assure sufficient residence time for the GAC in process. The upper part of this chamber was the incineration zone. Additional air and fuel were injected by the secondary burner to incinerate organics from both the reactivation and dryer zones. Temperatures across this chamber averaged 816°C (1500°F).

The furnace offgas was conducted from the reactivation/incineration chamber through a venturi scrubber and a tray scrubber and was discharged to the atmosphere by a stack through the roof. The breaching was equipped with a safety blow-out panel, which would release excessive pressures through a relief stack.

The third chamber, smallest in volume, was the dryer, which was separated from the reactivator/incinerator by a 316L stainless steel plate perfo-

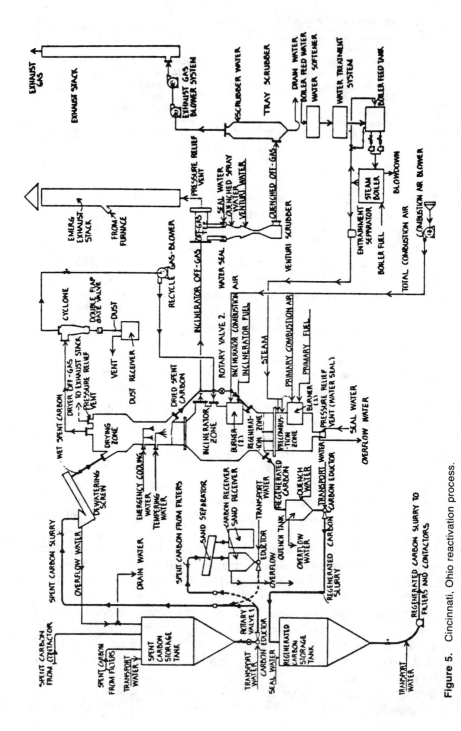

Figure 5. Cincinnati, Ohio reactivation process.

rated with a uniform pattern of holes to distribute the gas flow to attain uniform fluidization of the GAC bed above it. The gas flow through this chamber was induced by the suction of a fan, creating a negative pressure within the chamber. The gas flow went through a cyclone separator, which dropped out much of the suspended particulates (GAC fines). The blower then discharged into the incinerator zone, where the entrained volatiles driven off the GAC in the drying process and the remaining particulate matter were incinerated. The temperature within the dryer was approximately 150°C (300°F) and was controlled with cooling water sprays located above the GAC bed. The dryer chamber also had emergency water sprays and a blow-out panel-type pressure relief vent connected to the relief stack.

Also included were two cylindrically walled, conically bottomed tanks. Each had a capacity of 23.9 m³ (855 ft³), fabricated of 316L stainless steel, one for receiving and holding spent GAC prior to reactivation (ST-1) and the other to receive and hold reactivated GAC (ST-2) until it could be returned to the filter or contactor from which it came (Figure 3). Spent GAC was discharged from ST-1 through a variable speed rotary valve (RV-1). The rotational speed was controlled by the operator. This determined the rate of feed to the furnace. The GAC was moved from RV-1 by an eductor to either the sand separator or the dewatering screw.

A sand separator, which was an application of a standard piece of placer mining equipment, was utilized only when GAC from a filter was being reactivated. It was a vibrating table designed to separate materials through their difference in specific gravity. The sand was discharged to a portable dump hopper for disposal, and the GAC was discharged into a stainless steel vessel fitted with an eductor that transported the GAC to the dewatering screw.

The dewatering screw, which was an inclined screw conveyor, discharged wet GAC directly into the dryer section of the reactivator. The GAC traversed across the dryer plate and through a side pipe. It passed through a rotary valve (RV-2) and reentered the furnace at the reactivation zone. The rotary valve served to isolate the two pressure zones within the process. The GAC flowed across the reactivation bed to an outlet port and discharge chute to the quench tank. The discharge chute was submerged to prevent entry of excess oxygen into the process. An eductor at the bottom of the quench tank transported the GAC to ST-2. A rubber-lined slurry pump was used to transfer the reactivated GAC to a filter or a contactor.

Foremost of the support equipment was a Bristol microprocessor/controller and recording unit. This was housed in a controlled environment room within the main building. The instrumentation and control systems were designed to require only the minimum of attention by the operator. However, it was seldom that all of the automated systems were functional and the successful operation relied heavily upon the skills of the CWW operating and maintenance personnel.

A three-stage turbo blower supplied the combustion air for both the reactivator and incinerator burners. The recycled gas blower, which handled the dryer offgas and reinjected it into the incinerator section, also provided the gas movement required to fluidize the dryer bed. Particulate matter in the recycled gases was removed by a cyclone separator. Two blowers in series handled the furnace offgases, which also passed through a venturi scrubber and a tray-type scrubber. All parts of the various gas-handling apparatus that were in contact with the gas stream were fabricated of 316L stainless steel.

There was also a steam generator to provide the steam required in the process. It, like the furnace, could be fired by either gas or fuel oil.

Appurtenant to the reactivator were the fuel oil tank and two 0.2-L/sec (3.2-gpm) centrifugal pumps that supplied contactor effluent water for steam generation, quenching, and transporting reactivated GAC.

GAC Treated Process Water Piping and Pumps

Two 12.6-L/sec (200-gpm) pumps provided contactor effluent for the steam generator, tempering water, quenching water, and transporting reactivated GAC, thereby reducing exposure to organic-laden water.

The suction was initially tapped into the 40-cm (16-in.) contactor effluent line just before it left the contactor building. With all four contactors in service, the pumps became air-bound and lost their prime. The suction connection was moved to the drop leg as the 40-cm (16-in.) line entered the tunnel area, making a trap configuration and assuring a full line. Apparently, turbulence entrained sufficient air to cause the pumps to become air-bound.

Reliable service was achieved only by relocating the pumps to the floor of the stairwell that connects the building to the access tunnel (Figure 3). The 40-cm (16-in.) contactor effluent pipe was tapped adjacent to the pumps, providing a positive suction head with only two contactors in service. A 7.6-cm (3-in.) discharge header was routed to the north end of the building along the east wall.

GAC Transport Piping

A 7.6-cm (3-in.) carbon-in pipeline was connected to the top of each contactor tank from a central header that terminated near the center of the building. Similarly, a 7.6-cm (3-in.) carbon-out line ran from the bottom of each contactor to a header that terminated at a point approximately 30 cm (1 ft) below the carbon-in line. A third line that began outside the building provided a connection for unloading or loading GAC trucks and terminated on a 90° alignment to the carbon-in and carbon-out lines and at the same elevation above the floor. A quick-disconnect hose permitted various modes of interconnection between these lines and the GAC transport piping.

These lines were assembled with Victaulic connections to permit ease of rearrangement within the filter area and removal of section for clean-out in the event of plugging. All bends in the GAC transport system were of a long radial design in order to reduce frictional loss of GAC.

Adsorber (Contactor)/Reactivator Building

A standard modular building was selected that could house the contactor tanks, whose size and number were established, and could accommodate either an infrared electrical furnace (horizontal process) or a fluidized bed gas-fired furnace (vertical process). A tunnel connecting the proposed building to the existing filter plant was constructed, consisting of an adaptation of a large prefabricated culvert pipe.

Sampling Plan

Figure 6 summarizes the sampling points for the full-scale GAC system.

JEFFERSON PARISH, LOUISIANA

Facility Description

The East Jefferson Parish raw water intake is located at river mile 105.4 above the mouth of the Mississippi River.[2] At this site, raw water is pumped to four separate treatment plants with a total capacity of 265,000 m³/day (70 mgd). Finished drinking water was supplied to the GAC filtration and reactivation facility by the Permuttit III Plant as indicated in Figure 7, which depicts the general treatment train. The influent to the GAC pilot plant was finished drinking water that had been treated with fluosilicic acid for fluoridation, cationic polymers for clarification, chloramines for disinfection, and zinc sodium hexametaphosphate for corrosion control. The effluent of the GAC adsorbers was disinfected with free chlorine to provide sufficient in-line disinfection before the effluent reentered the distribution system. This particular process scheme was employed as a cost-cutting measure. Available funding was unable to support the original design, which included two gravity sand filters prior to GAC filtration to avoid predisinfection and a post-GAC filtration clearwell to allow sufficient contact time for chloramine disinfection.

The GAC pilot facility was composed of three 3785-m³/day (1-mgd) pressure contactors; storage tanks for spent, reactivated, and makeup GAC; a GAC transport system; and an infrared reactivation furnace with an infrared afterburner as indicated in Figure 8, which depicts the general layout of the facility.

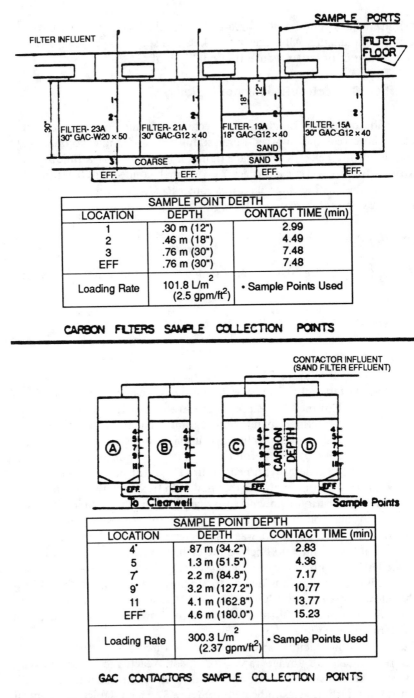

SAMPLE POINT DEPTH		
LOCATION	DEPTH	CONTACT TIME (min)
1	.30 m (12")	2.99
2	.46 m (18")	4.49
3	.76 m (30")	7.48
EFF	.76 m (30")	7.48
Loading Rate	101.8 L/m^2 (2.5 gpm/ft^2)	• Sample Points Used

CARBON FILTERS SAMPLE COLLECTION POINTS

SAMPLE POINT DEPTH		
LOCATION	DEPTH	CONTACT TIME (min)
4*	.87 m (34.2")	2.83
5	1.3 m (51.5")	4.36
7*	2.2 m (84.8")	7.17
9*	3.2 m (127.2")	10.77
11	4.1 m (162.8")	13.77
EFF*	4.6 m (180.0")	15.23
Loading Rate	300.3 L/m^2 (2.37 gpm/ft^2)	• Sample Points Used

GAC CONTACTORS SAMPLE COLLECTION POINTS

Figure 6. Cincinnati, Ohio full-scale GAC system sample point locations.

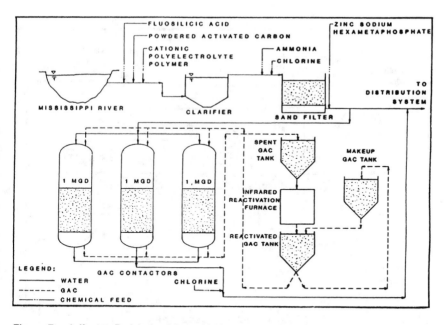

Figure 7. Jefferson Parish, Louisiana drinking water treatment train.

GAC Contactors and Storage Tanks

As indicated in Figure 9, each of the three 3785-m³/day (1-mgd) pressure contactors was 3.6 m (12 ft) in diameter by 9.6 m (31.5 ft) in height and designed for a maximum operating pressure of 4.92 kg/cm² (70 psi). Water entering each contactor was distributed by a 20.3-cm (8-in.)–diameter influent header with 10.2-cm (4-in.)-diameter laterals. The 316L stainless steel screen mesh surrounding the laterals was later abandoned due to plugging. The contactors contained approximately 5 m (16.5 ft) or 52.7 m³ (1860 ft³) of GAC, which with a hydraulic loading of approximately 244.2 L/min/m² (6 gpm/ft²) produced an EBCT of around 20 min. Water sampling laterals were located at the quarter height points of the GAC bed. These laterals were occasionally dislodged during bed expansion and were eventually replaced by well screen nozzles attached directly to the walls. GAC sampling ports were located at approximately 0.61-m (2-ft) intervals across the bed. Four GAC depth measuring ports were located on the top of each contactor for determining the volume of the GAC bed. The GAC bed was supported by a shallow cone underdrain plate that contained 212 316L stainless steel well screen nozzles. These nozzles were attached to the underdrain plate via "T" bolts, since the originally supplied "L" bolts allowed the nozzles to become cocked during backwashing and permitted GAC to penetrate the underdrain. The underdrain plate was reinforced by a series of I-beam and steel plate supports. The entire interior of each contactor, including the

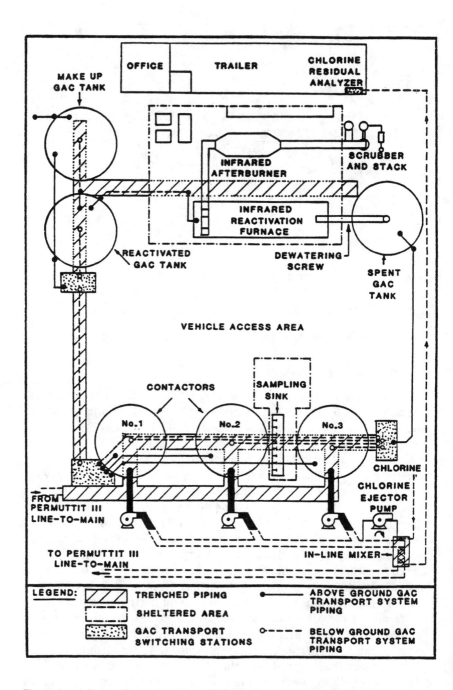

Figure 8. Jefferson Parish, Louisiana GAC filtration and reactivation facility layout.

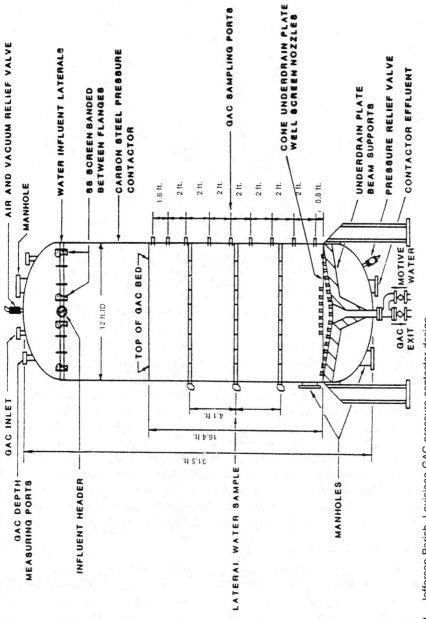

Figure 9. Jefferson Parish, Louisiana GAC pressure contactor design.

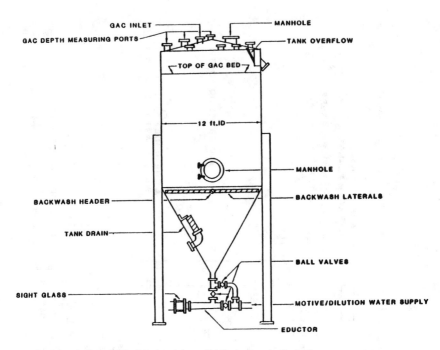

Figure 10.   Jefferson Parish, Louisiana GAC storage tank design.

influent header and laterals, was coated with Wisconsin Plasite 4020 paint to the thickness of 40 mils. This coating was selected through a coating evaluation study that examined the organics leaching, GAC abrasion, and static corrosion in moist GAC of eight coatings from six manufacturers.

The GAC storage tanks depicted in Figure 10 were 3.6-m (12-ft)-diameter nonpressure vessels designed to contain one contactor load of GAC. They were provided with a backwashing header 15.2 cm (6 in.) in diameter and laterals 10.2 cm (4 in.) in diameter in addition to a well screen drain to facilitate GAC volume measurements via the 10 measuring ports on the top of each tank. The bottom of the spent GAC storage tank was fitted with a rotary valve for GAC transport, while those of the reactivated and makeup storage tanks were fitted with a 6.4-cm (2.5-in.) eductor.

GAC Transport System

A GAC slurry system was employed to transport GAC throughout the facility, as indicated in Figure 11. The transport system was comprised of 3.2-cm (1.25-in.) and 10.2-cm (4-in.) steel pipe along with 3.2-cm (1.25-in.) and 6.4-cm (2.5-in.) 316L stainless steel eductors where vessel pressurization was not employed. The transport system also contained a rotary valve and a dewatering screw. (The latter was supplied as part of the reactivation system.) The GAC slurry portion of the transport system was driven by the

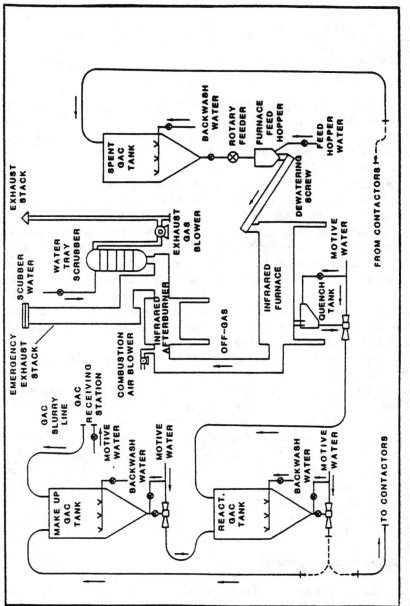

Figure 11. Jefferson Parish, Louisiana GAC transport and reactivation system.

Table 5.   GAC Transport System Operating Conditions

	Contactors	Furnace Quench Tank	Reactivated and Makeup Tanks
Eductor size, cm (in.)	–	3.18 (1.25)	6.36 (2.5)
Transfer line, cm (in.)	10.18 (4)	3.18 (1.25)	10.2 (4)
Motive water, L/min (gpm)	1892.5 (500)	75.7 (20)	511 (135)
Velocity, cm/sec (ft/sec)	396.2 (13)	274.3 (9)	152.4 (5)
GAC/water ratio, g/L (lb/gal)	0.9 (0.6)	0.3 (0.2)	1.4 (0.9)

3.9–4.2-kg/cm^2 (55–60-psi) line pressure from the Permuttit III line-to-main which supplied influent water to the GAC facility. GAC slurry switching stations were comprised of quick-connect hose couplings banded to a wire-reinforced rubber hose. In-line sight glasses were employed to visually estimate GAC slurry density.

Once a contactor was taken offline, the spent GAC was transferred to the spent GAC storage tank without the use of an eductor by the controlled pressurization of the contactor using the backwash flow rate control valve. Dilution water was added at the bottom of the contactor during the transfer at a rate of 1892.5 L/min (500 gpm). From the spent GAC storage tank, GAC was fed into the furnace feedhopper via a rotary valve. From the feedhopper, GAC was fed into the furnace via a dewatering screw, which reduced the moisture content of the GAC to approximately 50% by weight. Upon exiting the furnace, reactivated GAC was transported as a water slurry from the furnace quench tank to the reactivated GAC tank via a 3.2-cm (1.25-in.) eductor. Upon the completion of the reactivation of each contactor load of GAC, virgin makeup GAC was added to the reactivated GAC storage tank via the 6.4-cm (2.5-in.) eductor at the bottom of the makeup tank to replace the GAC lost during reactivation and transport. In a similar fashion, the GAC within the reactivated storage tank was then transferred to the next contactor to come off-line.

The operating conditions of the GAC slurry portion of the transport system are summarized in Table 5. Eductor and transfer line sizes as well as water flow rates were considerably varied, producing GAC water ratios of 0.3–1.4 g/L (0.2–0.9 lb/gal) and transfer line velocities of 152.4–396.2 cm/sec (5–13 ft/sec). No effort to optimize GAC slurry transport conditions was made other than to maintain the GAC slurry density at 1.5 g/L (1 lb/gal) or below.

GAC Reactivation System

The reactivation of spent GAC was accomplished using a 97,522-g/hr (215-lb/hr) Shirco infrared furnace. The system was composed of a feedhopper, a dewatering screw, a furnace module with a water quench tank, an afterburner with a combustion air blower, an offgas water pre-cooler and water tray scrubber, an exhaust draft blower with both normal and emergency bypass stacks, electrical power control centers for each temperature zone, and an automatic microprocessor-based control and alarm panel. The reactivation process involved sequential drying, volatilization and/or pyrolysis of absorbed organics, and reactivation with flue gas and steam. After the spent GAC was dewatered to approximately 50% moisture by weight by the dewatering screw, it was transported on a special alloy conveyor belt at a depth of 1.90 cm (0.75 in.) through the various temperature zones of the furnace. The drying zone of the furnace was typically maintained at 260–316°C (500 to 600°F) by residual heat from the pyrolysis zone, which was heated to 760°C (1400°F). The reactivation zone of the furnace was maintained at 927°C (1700°F). To provide a minimum amount of GAC agitation during the reactivation process, cakebrakers were located in the drying zone, in the center of the pyrolysis zone, and at the beginning of the reactivation zone. After 25 min of resident time within the furnace, the reactivated GAC was quenched with water and subsequently educted to the reactivated GAC storage tank. Changes in the reactivation conditions could be made, when required, by changing the reactivation zone temperature and the furnace residence time. The reactivation conditions were based upon thermogravimetric analysis performed by the manufacturer on a composite sample of spent GAC from Lot No. 1 each time this lot was reactivated. The offgases from the furnace were drawn through the afterburner, which was maintained at 1010°C (1850°F). The offgas residence time within the afterburner was approximately 0.3 min. Since the oxygen level within the furnace was kept to a minimum through the generation of steam in the drying zone, a small amount of combustion air was introduced prior to the afterburner to promote combustion. The afterburner offgas was quenched and scrubbed in a water tray scrubber before exiting the 10.2-cm (4-in.)–diameter stack with an average flow rate of 54.2 L/sec (115 scfm), a particle emission rate of approximately 1 g/hr, and a temperature of less than 49°C (120°F).

The infrared reactivation system was capable of manual, semiautomatic, and totally automatic operation. Automatic operation was provided by a microprocessor-based process/sequence controller. The microprocessor was capable of performing automatic shutdown of the entire system in the event of a major malfunction and of alerting the operator to a particular malfunction. During the first 11 reactivations, the reactivation system was operated in the semiautomatic mode, while problems associated with GAC bridging in the quench tank, the furnace feed chute level control probe, and the size of the quench tank eductor were resolved. Thereafter, the system was operated in the fully automatic mode, requiring very little operator attention. The furnace reactivated 20 contactor (adsorber) loads of GAC. The automatic control system was comprised of an alarm indicator panel with both visual and audible alarms; kilowatt meters and totalizers for the two zones of the furnace and the afterburner; a furnace belt speed control; a run time indicator; the microprocessor with a cathode ray tube (CRT) readout of all control parameters; an oxygen analyzer for the furnace and a combination oxygen/combustion analyzer for the afterburner; and strip chart recorders for the oxygen levels and the temperatures within the furnace and afterburner.

Several problems that developed in the reactivation system were essentially resolved during the first 11 reactivations. Originally, the furnace belt was continually off-track, prompting the installation of an automatic pneumatic belt alignment system. GAC bridging in the quench tank chute, which caused the quench tank to overflow, was rectified through the installation of a water spray within the chute. An excessive buildup of GAC in the quench tank was resolved by replacing the 2.54-cm (1-in.) quench tank eductor with a 3.18-cm (1.25-in.) eductor. The original furnace feed chute level control probe, which was insulated by a ceramic sleeve, frequently gave false high level readings and was replaced with a Teflon-coated probe. During the 14th reactivation, warping of the top of the quench tank chute caused GAC bridging in the quench tank. The chute was straightened and a length of 1.27-cm (0.5-in.) stainless steel bar stock was welded across the center of the chute. Also, the combustion air blower supplied was unnecessary due to the low level of combustibles adsorbed by GAC in drinking water applications.

While a number of problems with the reactivation furnace have been resolved, several problems remain unresolved. A buildup of GAC in the lower chamber of the furnace resulted in the furnace having to be shut down and cleaned out twice during each reactivation cycle, once in the middle of the cycle, and again at the end of the cycle. Each cleanout resulted in a furnace downtime of 6–8 hr. The GAC removed was relatively fine, with a mean particle diameter of approximately 0.5 mm, as compared to 1.2 mm for virgin GAC, and accounts for approximately 2% of the GAC loss observed across the reactivation furnace. Observations have indicated that these GAC particles are sticking inside the mesh of the furnace belt and are being released when the belt flexes over the drive drum and associated rollers. It is possible that this problem can be minimized or even eliminated by a redesign of the furnace belt weave.

Another remaining problem appears to be excessive glowbar usage in the infrared afterburner. After 11 reactivations, 15 glowbars have had to be replaced in the afterburner to maintain a temperature of 1010°C (1850°F), while none were replaced in the furnace, which has a pyrolysis zone temperature of 760°C (1400°F) and a reactivation zone temperature of 927°C (1700°F). While this may be attributable in part to the higher oxygen level in the afterburner (approximately 6–8% higher than that in the furnace), it would also appear that the afterburner was designed too small relative to the low level of combustibles loaded onto GAC in drinking water applications (as compared to wastewater applications). The afterburner was designed to take advantage of the spontaneous combustion of volatiles coming from the reactivation furnace to supply a significant portion of the heat necessary to maintain 1066°C (1950°F). Because of the low level of combustibles on the GAC in this application, the afterburner was continuously operated at about 100% of maximum power to maintain the desired temperature. The afterburner was equipped with nine glowbars (powered by a 75 kva power center) and was required to maintain 1010°C (1850°F). In contrast, the reactivation zone of the furnace was equipped with 12 glowbars powered by a 125 kva power center that cycled from approximately 0–100% of maximum power in order to maintain 927°C (1700°F).

Corrosion was a developing problem, particularly in the duct between the furnace and the afterburner. After 12 reactivations, the top of the duct connecting the furnace and the afterburner was observed to have corroded through in two places, just before and after the furnace oxygen analyzer. A total of six holes were found by placing smoke bombs inside the furnace. The two larger holes were 2.5 × 5.1 cm (1 × 2 in.) and 2.5 × 7.6 cm (1 × 3 in.) with the remaining holes, about 0.6 cm (0.25 in.) in diameter, spaced around the larger holes. These holes were temporarily sealed with silicone sealant.

MANCHESTER, NEW HAMPSHIRE

Manchester used a 500-lb (227-kg)-per-hour fluidized bed process (rated the same capacity as CWW's reactivator).[3] The furnace was divided into four temperature zones: drying, incineration, reactivation, and combustion. The furnace was 6.8 m (22.5 ft) tall and 2.1 m (7 ft) in diameter at its widest point. The 0.6-cm (0.25-in.) outer steel shell was covered with a cast-in-place refractory lining up to 0.36 m (14 in.) in thickness.

Spent GAC was hydraulically educted into a storage tank where the carbon was metered through a rotary valve to a dewatering feed screw. Dewatered GAC (50% moisture) was dried in the drying zone at approximately 150°C (300°F). Dried GAC passed from the drying zone through a discharge pipe and rotary metering valve to the reactivation zone of the furnace. Temperatures within the reactivation zone typically averaged about 816°C (1500°F).

In the reactivation zone, part of the adsorbate was volatilized from the GAC while the remaining pyrolyzed portion was removed by selective reaction with steam to form hydrogen and carbon monoxide. The adsorbate was pyrolyzed in the incineration zone directly above the reactivation bed. Excess gas from incineration was bypassed and mixed with offgas from the dryer, with the system gas treated in a three-stage venturi-type scrubber and water tray scrubber to remove particulate matter before discharge.

The combustion chamber (bottom section of furnace) utilized a balanced mixture of fuel and oxygen to control the quality of combustion gases in the fluidizing medium. Steam was also injected into this chamber to provide an additional flow of fluidizing gases and to temper combustion chamber temperature. The combustion chamber temperature was about 1038°C (1900°F). Reactivated GAC was discharged to a quench tank, where it was cooled before eduction to the reactivated GAC storage tank.

REFERENCES

1. Miller, R., and D. J. Hartman. "Feasibility Study of Granular Activated Carbon Adsorption and On-Site Regeneration," U.S. EPA Report 600/52-82-087A (November 1982).
2. Koffskey, W. E., N. V. Brodtmann, and B. W. Lykins, Jr. "Organic Contaminant Removal in Lower Mississippi River Drinking Water by Granular Activated Carbon Adsorption," U.S. EPA Report 600/2-83-032 (June 1983).
3. Kittredge, D., R. Beaurivage, and D. Paris. "Granular Activated Carbon Adsorption and Fluid-Bed Reactivation at Manchester, New Hampshire," U.S. EPA Report 600/2-83-104 (March 1984).

CHAPTER 7

Reactivation Performance

INTRODUCTION

Results of onsite reactivation conducted by Cincinnati, Jefferson Parish, and Manchester will be very valuable to other utilities contemplating the use of GAC. Therefore, some of the results of these reactivation studies and some of the problems encountered are discussed here.

As mentioned earlier, Cincinnati and Manchester used a 227-kg (500-lb)-per-hour fluidized bed process (Figure 1). The furnace was divided into four temperature zones: drying, incineration, reactivation, and combustion. The furnace was 6.7 m (22.5 ft) tall and 2.1 m (7 ft) in diameter at its widest point. The 0.6-cm (0.25-in.) outer steel shell was covered with a cast-in-place refractory lining up to 0.36 m (14 in.) in thickness. Spent GAC was hydraulically educted into a storage tank where the carbon was metered through a rotary valve to a dewatering feed screw. Dewatered GAC (50% moisture) was dried in the drying zone at approximately 150°C (300°F). Dried GAC was moved from the drying zone through a discharge pipe and rotary metering valve to the reactivation zone of the furnace. The temperatures within the reactivation zone typically averaged about 816°C (1500°F). In the reactivation zone, part of the adsorbate was volatilized from the GAC, while the remaining pyrolyzed portion was removed by selective reaction with steam to form hydrogen and carbon monoxide. The adsorbate was pyrolyzed in the incineration zone directly above the reactivation bed. Excess gas from the incineration zone was bypassed, mixed with offgas from the dryer, and with the system gas treated in a three-stage venturi-type scrubber and water tray scrubber to remove particulate matter before discharge. The combustion chamber (bottom section of furnace) utilized a

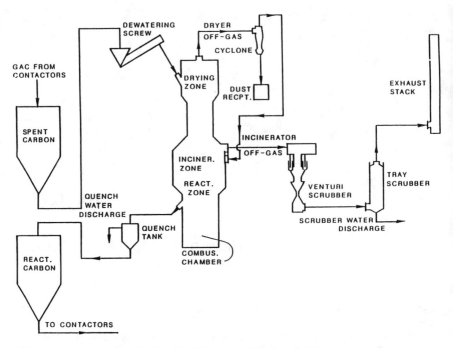

Figure 1. Schematic of fluidized bed furnace at Cincinnati, Ohio.

balanced mixture of fuel and oxygen to control the quality of combustion gases in the fluidizing medium. Steam was also injected into this chamber to provide an additional flow of fluidizing gases and to temper combustion chamber temperatures. The combustion chamber temperature was about 1038°C (1900°F). Reactivated GAC was discharged to a quench tank, where it was cooled before eduction to the reactivated GAC storage tank.

An infrared furnace was used onsite at Passaic and offsite at Evansville for reactivation of the spent GAC (Figure 2). This 45.4-kg (100-lb)-per-hour unit was 1.2 m (4 ft) wide by 6.1 m (20 ft) long. The GAC was dewatered during conveyance to the furnace to approximately 50% moisture. At the furnace, it dropped onto a woven wire conveyor belt and was leveled into a layer approximately 1.9 cm (0.75 in.) thick. The conveyor belt moved the GAC underneath infrared heating elements that provided the energy necessary to dry the carbon, drive off adsorbed compounds, and restore pore structure. Process temperatures were controlled in zones and typically ranged from 649°C (1200°F) in the drying zone to about 899°C (1650°F) in the reactivation zone. The furnace exhaust temperature was about 649°C (1200°F) and the wet-scrubbed exhaust gas discharged to the atmosphere was about 38°C (100°F). Residence time of the GAC in the furnace was from 20 to 30 min.

Jefferson Parish also used infrared reactivation. Because of the magni-

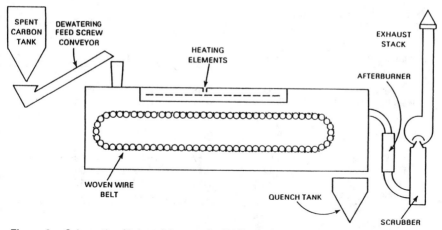

Figure 2.   Schematic of infrared furnace for GAC reactivation.

tude of the Jefferson Parish effort, it will be discussed in a separate section.

GAC PROPERTY MEASUREMENTS

As an indication that proper reactivation of spent GAC had occurred, several GAC properties were evaluated. One of the easier properties to determine that is widely accepted as an immediate indication of the efficiency of the reactivation process is the apparent density.[1] (Apparent density is the weight per unit volume of a homogeneous activated carbon. To assure uniform packing during measurement, a vibrating trough is used to fill the measuring device.) Data for this property for both furnace types show that the spent GAC was essentially restored to its virgin state (Table 1).

The general apparent density trend for the fluidized bed and infrared furnace yielded comparable values for subsequent reactivations when compared to the virgin GAC. A more vigorous method is used to reactivate the GAC in the fluidized bed furnace than in the infrared furnace, leading to a concern that reduced apparent density and mean particle diameter may occur. (Mean or average particle diameter is a weighted average diameter of a granular carbon. A screen analysis is run and the average particle diameter calculated by multiplying the weight of each fraction by its average diameter, adding the products, and dividing by the total weight of the sample. The average diameter of each fraction is taken as the size midway between the sieve opening through which the fraction has passed and the sieve opening on which the fraction was retained.) If reduction in mean particle diameter occurs, filtration rates are effected by increased filter head loss and possible media loss during backwashing. Establishing the proper conditions

Table 1. Apparent Density of Reactivated GAC

	Virgin GAC (g/mL)	Subsequent Reactivations			
Furnace Type		1 (g/mL)	2 (g/mL)	3 (g/mL)	4 (g/mL)
Fluidized bed					
Cincinnati[a]	0.48	0.48	0.47	0.47	0.52
Manchester[b]	0.61	0.59	0.58	0.56	—
Infrared					
Evansville[c]	0.43	0.45	0.40	0.44	—
Passaic					
WVW 12 × 40	0.58	0.57	0.58	0.59	
F400	0.47	0.49	0.49	0.50	—
HD 10 × 30	0.44	0.43	0.43	0.43	
Jefferson Parish[a]	0.54	—	0.54	0.57	—

[a]WVG 12 × 40.
[b]WVW 8 × 30.
[c]HD 10 × 30.

for reactivation is essential to remove the adsorbate without damaging the original GAC pore structure. An example of the effect on the mean particle diameter during subsequent reactivations for both furnace types is shown in Table 2. These data show that the mean particle diameter was reduced regardless of furnace type or GAC type.

Other properties seem to follow the same pattern whereby subsequently reactivated GAC was essentially restored to a state that was somewhat different than the original virgin carbon. This, however, does not mean that the reactivated GAC performed less efficiently than the virgin GAC. In many cases, the opposite was observed, as will be shown later. An indication of gradual degradation of adsorptive capacity after subsequent reactivations can be shown by a correlation of the cumulative surface area and iodine number. A ratio of about one indicates that the GAC pores were opened and maintained.[2] These ratios are presented in Table 3 and show that the Cincinnati and Evansville reactivations indicate an unlikelihood of

Table 2. Mean Particle Diameter of Subsequently Reactivated GAC

	Virgin GAC (g/mL)	Subsequent Reactivations			
Furnace Type		1 (g/mL)	2 (g/mL)	3 (g/mL)	4 (g/mL)
Fluidized bed					
Cincinnati[a]	1.25	1.18	0.95	0.86	1.02
Manchester[b]	1.42	1.37	1.38	1.24	—
Infrared					
Evansville[c]	1.29	1.14	1.12	0.94	—
Jefferson Parish[a]	1.20	—	1.12	1.11	—

[a]WVG 12 × 40.
[b]WVW 8 × 30.
[c]HD 10 × 30.

Table 3. BET Surface Area[a]/Iodine Number[b] Ratios of Subsequently Reactivated GAC

Furnace Type	Virgin GAC	Subsequent Reactivations			
		1	2	3	4
Fluidized bed					
Cincinnati[c]	0.93	1.01	1.01	0.96	0.89
Manchester[d]	0.99	0.90	0.78	0.70	—
Infrared					
Evansville[e]	1.08	1.06	1.08	1.03	—
Jefferson Parish[c]	1.02	1.02	0.87	0.92	—

[a]Amount of surface per unit weight of carbon. The surface area of activated carbon is usually determined from the nitrogen adsorption isotherm by the Braunauer, Emmett, and Teller (BET) Method. Surface area is usually expressed in square meters per gram of carbon.
[b]The iodine number is the milligrams of iodine adsorbed by 1 g of carbon at an equilibrium filtrate concentration of 0.02 N iodine. It is measured by contacting a single sample of carbon with an iodine solution and extrapolating to 0.02 N by an assumed isotherm slope. The iodine number can be correlated with ability to adsorb low-molecular-weight substances.
[c]WVG 12 × 40.
[d]WVW 8 × 30.
[e]HD 10 × 30.

any adsorptive capacity degradation. However, Manchester and Jefferson Parish show decreasing ratios, which indicates the possibility of reduced adsorptive capacity.

GAC OPERATIONAL LOSSES

From an economic standpoint, the most important GAC operational criterion to consider is carbon loss resulting from reactivation and transport. There are two basic methods used in determining GAC losses. One method is mass balance or percent loss by weight, and the other is volumetric or percent loss by volume. The mass balance measurements were questionable because of the difficulty in accounting for losses attributed to carbon burnoff in the reactivator and during backwashing. Also, assumptions had to be made to account for bulk density differences and particle size changes.

Therefore, the volumetric method of measuring GAC losses was used. As a matter of practice, water utilities are primarily concerned with the volumetric levels of GAC in filter operation and not the mass.

To characterize the losses around the reactivation system, three different measured losses were determined on the fluidized unit (filter to filter, reactivation only, and transport) and two on the infrared unit (filter to filter and reactivation). Transport losses were determined by subtraction for the infrared unit. Several measurements were taken during each of many reactivation cycles. The average of these losses for adsorbers and filters contain-

Table 4. Average Percent GAC Losses, by Volume

Furnace Type	Transport	Reactivation	Total
Fluidized bed			
Cincinnati[a]	3.1	12.2	15.3
Manchester[b]	2.0	10.9	12.9
Infrared			
Evansville[c]	3.0	5.0	8.0
Jefferson Parish[a]	2.0	7.0	9.0

[a]WVG 12 × 40.
[b]WVW 8 × 30.
[c]HD 10 × 30.

ing GAC only (no sand) are shown in Table 4. Because of the measurement techniques that had to be used, some data fluctuations were noted and the extreme values were discarded.

At the Cincinnati location, GAC on a sand base was also evaluated. Total losses for this mode of operation were higher than for GAC only (18.5% vs 15.3%). Although a sand separator placed before the furnace removed most of the sand, some did enter the furnace, causing less-than-optimum furnace operation. This problem, along with longer transport, GAC removal by shovel, and sand base inconsistency, caused the losses to be greater.

REACTIVATION FURNACE DISCHARGE

Because various organic compounds are adsorbed by GAC, there is a concern about the possible gaseous or liquid discharge of these organics or other organics formed within the furnace. These results are summarized in this section but will be discussed in more detail in a later section. During the Cincinnati study, an analysis for methane equivalent was obtained for the stack discharge. How this parameter related to specific organic compounds is unclear, although the Southwest Ohio Air Pollution Authority stated that the emission quality was acceptable. A summary of the offgas analyses for Cincinnati is presented in Table 5.

After the original GAC research effort was completed at Cincinnati,

Table 5. Summary of Average Data for Reactivation Furnace Stack Analyses, Cincinnati, Ohio

	GAC Adsorber, g/hr (lb/hr)	GAC Filter, g/hr (lb/hr)
Particulate		
Filterable	4.5 (0.01)	4.5 (0.01)
Condensable	19.5 (0.04)	13.6 (0.03)
Methane equivalent[a]	131.7 (0.29)	81.7 (0.18)
Nitrogen oxide	163.4 (0.36)	202.5 (0.45)

[a]Total gaseous nonmethane organics is expressed as methane (CH_4) equivalent.

another evaluation of the GAC furnace discharge was initiated. This evaluation was to determine if tetrachlorodibenzo-*p*-dioxins (TCDDs) or tetrachlorodibenzo furans (TCDFs) were present in any of the furnace effluent streams.[3] The effluent streams sampled included the following:

1. stack exhaust (gases and particulate) after passing through the venturi and tray scrubbers
2. carbon fines removed from the drying zone offgas by the cyclone
3. combined liquid discharge from the venturi and tray scrubbers
4. overflow water from the reactivated GAC quench tank
5. spent GAC feed to the reactivation unit

The GAC used for this study had been in service for about one year and was stored in the adsorber for about six months before reactivation. TCDDs and TCDFs were detected in low concentrations. In nearly all cases, the TCDDs and TCDFs were attached to the particulate material. Comparison of TCDD concentrations detected to acceptable emission discharges from municipal waste combusters shows that the GAC reactivation stack discharge was about one order of magnitude less.[4]

The health concern for this type of discharge was addressed in an EPA document that stated, "Present estimates of potential TCDD emission from municipal waste combusters suggest that such releases do not present a public health hazard for residents living in the vicinity of the plants."[5] A more detailed discussion of both Jefferson Parish and Cincinnati and a risk assessment of the reactivation furnace stack emission discharge at Cincinnati will be discussed later.

REACTIVATION FURNACE OPERATION

Infrared furnace manufacturers have claimed ease of operation with quick startup and shutdown capabilities. Because an extensive evaluation of infrared furnace operation had not been done, these claims were not endorsed or refuted. A complete evaluation of operation will be done during a second Jefferson Parish research project. However, there have been some indications of various problems. These include belt stretching, lamp breakage, 304 stainless steel components deteriorating, and inconsistent leveling of GAC on the belt.

An extensive evaluation was done of the fluidized bed furnace at two locations (Cincinnati and Manchester). There were numerous problems with maintaining consistent furnace operation. Some of these problems were related to operator training and lack of a sufficient spare parts inventory and some were caused by inherent deficiencies in construction of the furnace. By implementing steps to improve operation and correcting major deficiencies, furnace uptime was improved from a low of 30% to about 80%.

This information is presented so that the reader is aware that the reactivation furnace operation can be labor- and time-intensive. Proper operator training is a must; operational efficiency improves with experience.

VIRGIN VS REACTIVATED *GAC*

As mentioned previously, spent virgin GAC was subsequently reactivated during research at five project sites: onsite at Cincinnati, Manchester, Passaic, and Jefferson Parish, and offsite at Evansville. Although two different furnace types were used (fluidized bed at Cincinnati and Manchester and infrared at Passaic, Evansville, and Jefferson Parish), the same performance criterion was used; namely, return the GAC back to essentially virgin state as determined by its properties.

The ideal approach for determining the effects of reactivation is to determine the adsorbent performance of subsequently reactivated GAC as compared to virgin GAC, using parallel systems so that the same influent is applied to both. This was only possible in Evansville. The other locations evaluated their systems at different periods of time with GAC influent variations such as temperature and water quality. When looking at performance, however, percent removal and organic loading were used for comparative purposes to normalize the varying influent concentrations and GAC weights.

For consistency, only posttreatment adsorbers are compared. In Cincinnati, the TOC influent concentration varied from 1.8 to 2.6 mg/L. Virgin GAC received 1.9 mg/L TOC; once- and twice-reactivated GAC received 2.6 and 1.8 mg/L, respectively. A comparison of the percent TOC removal in Figure 3 shows that all systems from virgin through twice-reactivated GAC performed essentially the same (within 15%) for up to 60 days of operation. The loading on these GAC systems, shown in Figure 4, also indicates that the three systems were comparable (100% removal would produce a 45° slope).

The closeness of the GAC performance was also reflected in the GAC properties as shown in Table 6. With the molasses number* increasing and the BET surface area decreasing, indications are that the pore structure had been changed to reflect the destruction of smaller pores with the creation of larger pores. This change, however, did not seem to effect the performance of the GAC for TOC removal and only slightly changed the mean particle diameter.

---

*The molasses number is calculated from the ratio of the optical densities of the filtrate of a molasses solution treated with a standard activated carbon and the activated carbon in question. The Decolorizing Index Unit (DIU) can be correlated with the capacity to adsorb many high-molecular-weight substances.

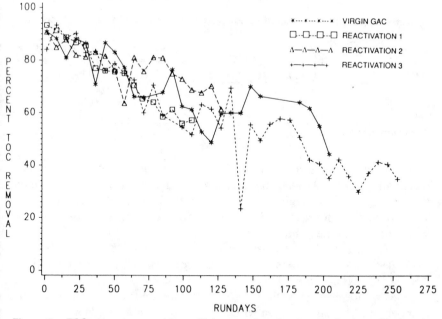

Figure 3.   TOC percent removal for multiple runs of posttreatment adsorbers, Cincinnati, Ohio.

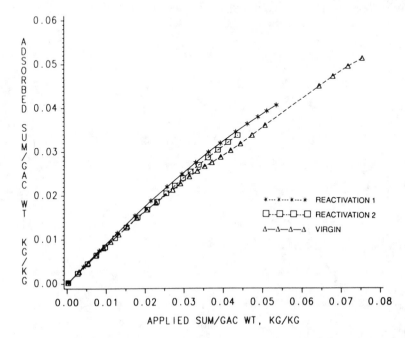

Figure 4.   TOC adsorbed per GAC weight for multiple runs of posttreatment adsorbers, Cincinnati, Ohio.

Table 6.   Comparison of Properties for Subsequently Reactivated Postadsorption GAC at Cincinnati, Ohio (Fluid Bed)

GAC State	Iodine Number (mg/g)	BET Surface Area (m²/g)	Molasses Number (DIU)[a]	Apparent Density (g/mL)	Particle Diameter (mm)
Virgin	1128	1070	9.6	0.506	1.10
Once-reactivated	1030	1046	10.7	0.508	1.04
Twice-reactivated	1096	1013	12.6	0.495	1.02

[a]Decolorizing index unit.

At Manchester, the spent virgin GAC was reactivated twice. With each reactivation, this carbon appeared to improve in performance, as shown by the loading curves in Figure 5. This performance improvement, however, seems to have been produced by over-reactivation of the GAC as indicated by an increase in the iodine number. Furnace conditions for reactivation of the GAC were determined by the quickest method possible (apparent density). Because of the inherent inaccuracies of this test, furnace conditions were not always appropriately established. Table 7 shows the GAC property changes with each reactivation. Of special note are the increases in iodine number and BET surface area with a corresponding drop in apparent density. The mean particle diameter remained essentially the same, although it would have been expected to decrease.

With less vigorous reactivation at Passaic (infrared furnace), all three

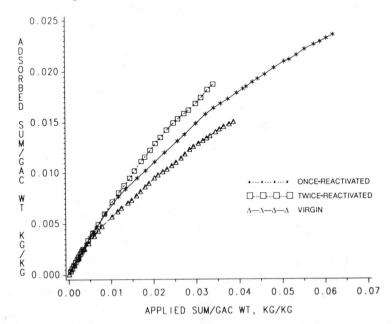

Figure 5.   TOC loading on virgin and subsequently reactivated GAC, Manchester, New Hampshire.

Table 7. Properties for Virgin and Subsequently Reactivated GAC at Manchester, New Hampshire (Fluid Bed)

GAC State	Iodine Number (mg/g)	BET Surface Area (m²/g)	Molasses Number (DIU)[a]	Apparent Density (g/mL)	Particle Diameter (mm)
Virgin	795	789	3.2	0.614	1.40
Once-reactivated	970	816	5.2	0.596	1.30
Twice-reactivated	1203	892	4.4	0.573	1.40

[a]Decolorizing index unit.

types of GAC evaluated showed that the virgin GAC performed better than the subsequently reactivated GACs as indicated by the TOC performance curves (Figure 6). Although the properties are different for the various GACs, Table 8 shows that these properties are essentially restored to those of virgin carbon.

At Evansville the same influent water was applied to two contactors. One always contained virgin GAC and the other subsequently reactivated GAC. Figure 7 shows that for three reactivations the carbon in both contactors removed TOC essentially equally with the reactivated GAC, removing slightly more than the virgin GAC. TOC loading on the two systems was also about the same for each reactivation, with slightly more adsorbed by the reactivated GAC (Figures 8 and 9).

A trend appeared to develop for the iodine number and molasses decolorizing index with the data showing that the spent GAC was restored to essentially virgin state. The iodine number increased from 617 to 650 mg/g and the molasses decolorizing index decreased from 18.5 to 13.7 (Table 9). The iodine number and cumulative surface area correlated almost exactly, indicating that the pores were opened and maintained so that there was an unlikelihood of any gradual degradation of adsorptive capacity after subse-

Table 8. Properties of Virgin and Subsequently Reactivated GAC at Passaic Valley, New Jersey (Infrared)

GAC State	Iodine Number (mg/g)	Apparent Density (g/mL)	Effective Size (mm)
HD 10 × 30			
Virgin	488	0.42	0.86
Once-reactivated	581	0.42	0.78
Twice-reactivated	617	0.42	0.72
F400			
Virgin	890	0.49	0.74
Once-reactivated	923	0.49	0.75
Twice-reactivated	882	0.51	0.67
WVW 12 × 40			
Virgin	761	0.57	0.76
Once-reactivated	807	0.58	0.71
Twice-reactivated	758	0.59	0.64

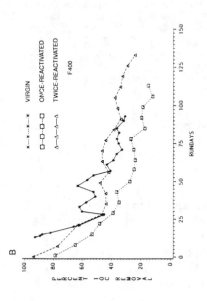

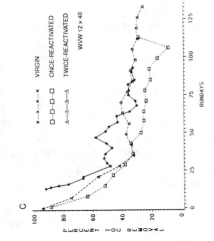

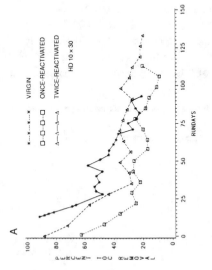

Figure 6.   TOC removal at Passaic, New Jersey.

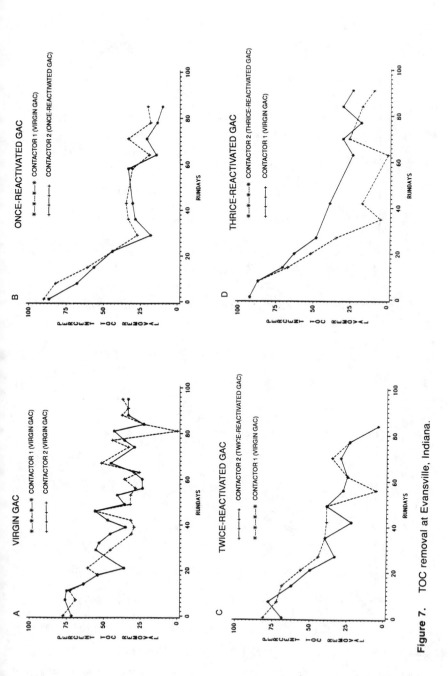

Figure 7. TOC removal at Evansville, Indiana.

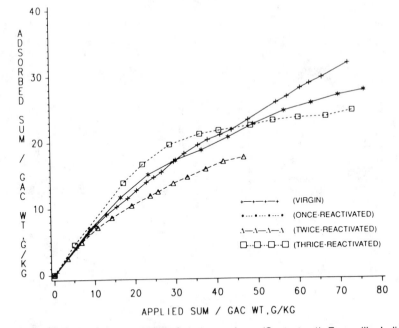

Figure 8. TOC loading for virgin GAC during each run (Contactor 1), Evansville, Indiana.

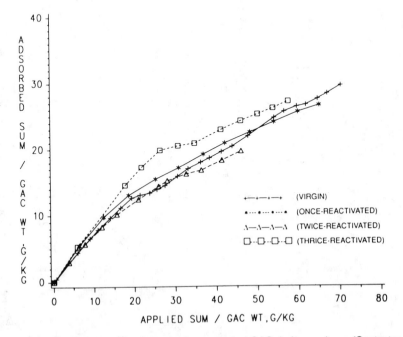

Figure 9. TOC loading for subsequently reactivated GAC during each run (Contactor 2), Evansville, Indiana.

Table 9. Properties for Virgin and Subsequently Reactivated GAC at Evansville, Indiana

GAC State	Iodine Number (mg/g)	BET Surface Area (m²/g)	Molasses Number (DIU)[a]	Apparent Density (g/mL)	Particle Diameter (mm)
Virgin	617	597	18.5	0.427	1.29
Once-reactivated	638	675	17.8	0.447	1.14
Twice-reactivated	596	644	14.7	0.403	1.12
Thrice-reactivated	650	688	13.7	0.440	0.94

[a]Decolorizing index unit.

quent reactivations. The ratios for BET surface area/iodine number for an average of virgin GAC and three subsequent reactivations were 1.08, 1.06, 1.08, and 1.03, respectively.

GAC morphology data for the first reactivation of GAC Lot 1 at Jefferson Parish were determined to be invalid due to an excessive number of fines found in the sample, with 42% of the sample passing a 40-mesh screen. The presence of these fines decreased the apparent density and the mean particle diameter to 0.43 g/mL and 0.56 mm, respectively, while increasing the iodine number to 1070, thus producing a somewhat distorted view of reactivation quality. After modifying the sampling location to obtain a more representative sample and after replacing only 11.6% of the bed with virgin GAC, more accurate GAC morphology data were obtained for the second reactivation, as indicated in Table 10.

Perhaps a better indicator of reactivation quality is the comparison of loading data between virgin and reactivated GAC. In Figure 10, the TOC loading of GAC Lot 1 is compared from virgin through the third reactivation. These curves are essentially identical up until 30-40 g/kg of applied TOC, at which point some reduction in reactivated GAC performance was observed. However, this was also the point at which "steady state" was reached, during which time factors other than adsorption, such as microbiological degradation, became more predominant. Figure 11 indicates that the TOC removal was roughly equivalent through the third reactivation.

Exhibiting somewhat of an opposite trend in Figure 12, the loading data for instantaneous trihalomethanes shows an increased performance of reac-

Table 10. GAC Morphology at Jefferson Parish, Louisiana

	Virgin	Reactivation 2nd	Reactivation 3rd
Apparent density (g/mL)	0.54	0.54	0.57
Iodine number (mg/g)	872	897	777
Molasses number	237	224	220
Mean particle diameter (mm)	1.20	1.12	1.11
Abrasion number	77.1	77.4	77.0
Ash (%)	8.9	10.0	8.7
BET surface area (m²/g)	892	777	714

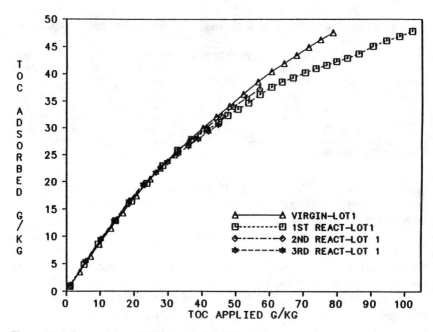

Figure 10.   Comparison of TOC loading on virgin GAC through the third reactivation.

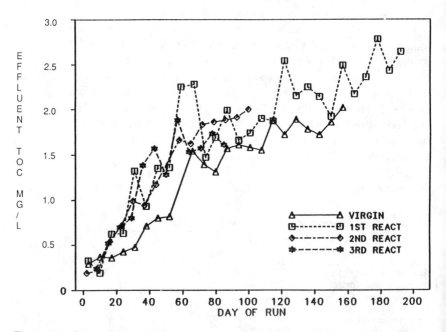

Figure 11.   Comparison of TOC effluent concentrations for virgin GAC through the third reactivation.

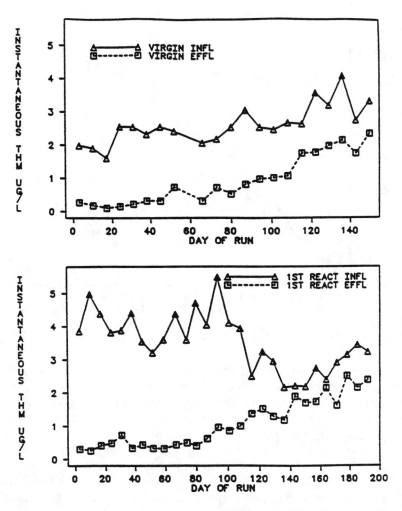

Figure 12. Comparison of the effluent concentrations for virgin GAC and once-reactivated GAC.

tivated GAC over virgin GAC. However, these loading data were generated from some relatively low THM concentrations resulting from chloramine disinfection with average influent and effluent levels of 3.5 and 1.5 μg/L, respectively. At these concentrations and without any apparent upset in the ammonia feed of the normal treatment plant, it took approximately six months for the instantaneous THMs to reach saturation, as indicated in Figure 13. This was somewhat unusual, as all prior full-scale filter and pilot column data[6,7] had indicated a period of about three months to saturation. This extended GAC life relative to THM removal appears to have resulted from a relatively constant influent concentration which in turn resulted

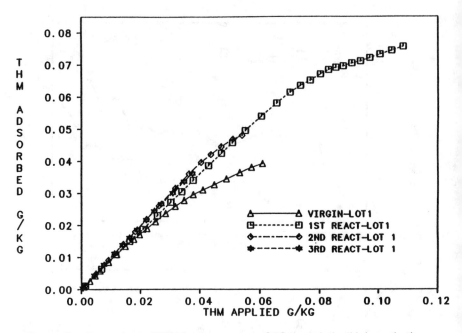

Figure 13.   Comparison of THM loading on virgin GAC through the third reactivation.

from a steady ammonia feed. Thus, no "spill event," as described by Neukrug et al., was created by a fluctuating ammonia feed resulting in a premature breakthrough of THMs.[8] The THM formation potential data collected have not been presented because they are still being evaluated to determine the effects of the presence of varying amounts of ammonia in the effluent of the GAC contactors. This resulted in the formation of varying amounts of chloramines that significantly suppressed the formation of THMs.

The TOX loading curves presented in Figure 14 are essentially identical for all three reactivations, indicating an identical performance of virgin and reactivated GAC relative to TOX removal. This was the case, even though once-reactivated GAC apparently reached a "steady state" condition at approximately 30 $\mu g/L$, as indicated in Figure 15. The TOX influent concentration ranged from 80 to 180 $\mu g/L$, with the effluent concentration ranging from 20 to 30 $\mu g/L$. After six months of operation, TOX removal efficiencies of approximately 90% were still being observed for both virgin and reactivated GAC.

BY-PRODUCTS OF REACTIVATION

Associated with reactivation is the potential of producing by-products from the precursor materials or organics adsorbed on the carbon. These by-products could consist of polychlorinated dibenzodioxins (PCDDs) and

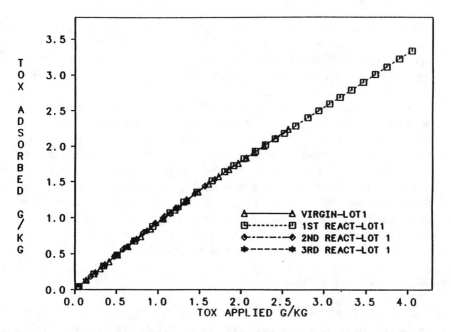

Figure 14. Comparison of TOC loading on virgin GAC through the third reactivation.

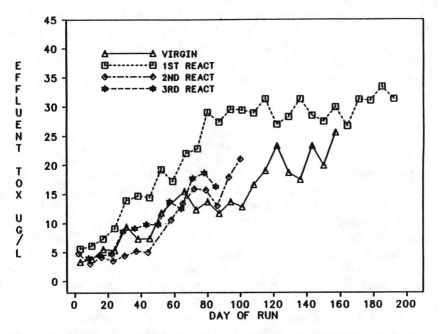

Figure 15. Comparison of TOX effluent concentrations for virgin GAC through the third reactivation.

polychlorinated dibenzofurans (PCDFs), depending on the precursors available for reaction in the reactivation furnaces. A major concern was that during thermal reactivation of spent GAC, organic compounds, including PCDDs and PCDFs, could be released into effluent streams that enter the environment.

Previous research has shown that PCDDs and PCDFs can chemisorb to particulate matter during combustion and be discharged into the atmosphere or emitted in a gaseous state. Atmospheric emissions of PCDDs/PCDFs from municipal incinerators, black-liquor recovery boilers, secondary copper recovery smelters, sewage sludge incinerators, and mine reclamation furnaces have been measured.[9-23] Therefore, it seemed likely that PCDDs and PCDFs could be emitted into the air during GAC reactivation. This likelihood, coupled with concern over possible adverse effects on public health, led to the stack testing of the two types of GAC reactivation furnaces evaluated for use at municipal waterworks.

GAC Furnace Operation

Although both Cincinnati and Manchester had fluidized bed furnaces, only Cincinnati was tested for emissions of PCDDs and PCDFs. Offgas sampling was performed at Manchester, but the work was completed before it became apparent to the investigators that PCDDs and PCDFs might be contained in the offgases. The furnace at Jefferson Parish is an infrared furnace, and the PCDD and PCDF testing performed at Jefferson Parish was similar to that performed in Cincinnati.

Figures 16, 17, and 18 show schematics for both fluid bed and infrared furnaces. These figures also show various points where sampling was conducted.

Sampling

Various effluent streams from the fluidized bed and infrared furnace were sampled during GAC reactivation. All sampling procedures, analytical procedures, equipment calibration, sample custody, data validation, quality control checks, performance and system audits, etc. were addressed in an acceptable comprehensive quality assurance plan that was instituted prior to evaluating the GAC furnace.[21,22] Stack, afterburner, and heat recuperator effluents were sampled using a modified Method 5 train shown schematically in Figure 19.[23,24] Stack sampling was performed isokinetically at a single point in the stack in a region of average velocity to obtain particulate and gas emissions. Grab samples were collected at 30-min intervals for spent carbon, reactivated carbon, cyclone fines, scrubber water, and quench water. The grab samples collected for each stream during a test were combined and homogenized to provide a composite sample for analysis.

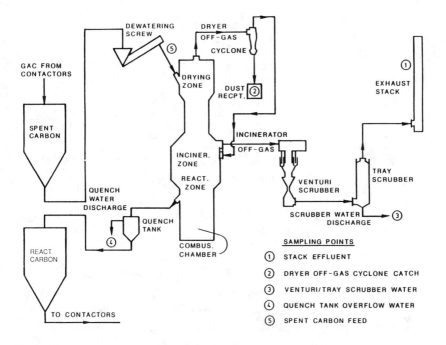

Figure 16. Phase 1 fluidized bed carbon reactivation system, Cincinnati Water Works.

Fluidized Bed Furnace

Evaluation of the fluidized bed furnace took place in two different operational phases. In Phase 1, chlorine was added to the water prior to GAC adsorption, interacting with precursor material and producing chlorinated products. During Phase 1, four tests were performed over a three-day period to determine the concentration of TCDDs and TCDFs in the reactivator effluent streams. These tests sampled the reactivator stack gas and particulates, dryer offgas cyclone particulates, scrubber and quench water, and spent carbon feed during reactivation of spent carbon that had been used approximately one year to treat Cincinnati water. Table 11 summarizes the sampling points and number of samples for each phase. During one test, the natural gas was sampled for polychlorinated biphenyls (PCBs).

In Phase 2, the addition of chlorine was eliminated prior to GAC adsorption. In Phase 2, an external afterburner was placed in the reactivation system (Figure 18). Seven tests were performed over a two-week period during Phase 2. Samples were collected for TCDD and TCDF analysis in all effluent streams and for other PCDDs and PCDFs in a preselected number of samples. Sampling was done for effluent streams identified in Table 11 during reactivation of carbon that had been used approximately 200 days to treat unchlorinated Cincinnati drinking water.

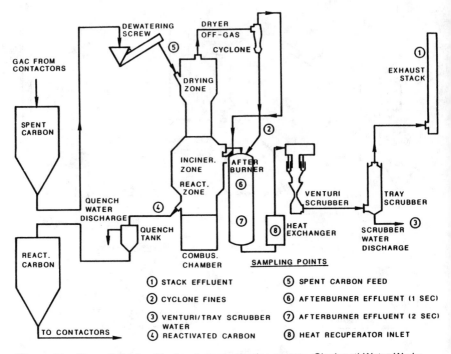

Figure 17. Phase 2 fluidized bed carbon reactivation system, Cincinnati Water Works.

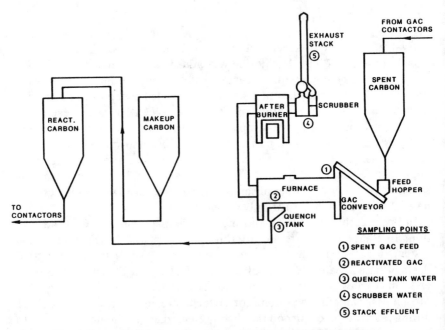

Figure 18. Jefferson Parish, Louisiana infrared carbon reactivation system.

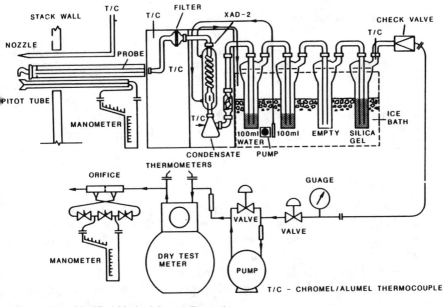

Figure 19. Modified Method 5 sampling train.

Table 11. Fluidized Bed Reactivator Sampling Points at Cincinnati Water Works

	Phase 1		Phase 2	
Sampling Points	Sample Location	Number of Samples	Sample Location	Number of Samples
Stack effluent	X	4	X	2
Dryer offgas cyclone catch	X	4		
Cyclone fines			X	2
Venturi/tray scrubber water	X		X	2
Quench tank water	X	4		
Reactivated carbon			X	1
Spent carbon feed	X	4	X	1
Afterburner effluent (1 sec)			X	1
Afterburner effluent (2 sec)			X	2
Heat recuperator inlet			X	4
Fuel for furnace	X	1	X	2

Table 12. Infrared Furnace Sampling Points

| Sampling Points | Virgin GAC | | Spent GAC | |
	Sample Location	Number of Samples	Sample Location	Number of Samples
Virgin carbon	X	3		
Spent carbon			X	3
Reactivated carbon	X	3	X	3
Quench tank water	X	3	X	3
Scrubber water	X	3	X	3
Stack effluent	X	3	X	3

Infrared Furnace

At Jefferson Parish, a batch of virgin carbon was reactivated to determine if any TCDD or TCDF could be detected. This provided a control for the subsequent evaluation whereby the same carbon was exposed to chloraminated water until exhaustion. Samples were collected over a three-day period for three tests for TCDD and TCDF analysis for the effluent streams shown in Table 12. A batch of spent carbon that had been used for approximately 156 days to treat chloraminated drinking water at Jefferson Parish was sampled during reactivation. Three tests were run for PCDDs and PCDFs over a three-day period during this reactivation for the furnace effluent streams shown in Table 12.

Analysis

PCDD and PCDF analyses of the samples involved three basic steps: (1) extraction of the PCDDs and PCDFs from the sample matrix, (2) sample cleanup to remove coextracted interferences, and (3) analysis of PCDDs and PCDFs by high-resolution gas chromatography/high-resolution mass spectrometry (HRGC/HRMS). Prior to extraction, each sample was spiked with 2,3,7,8-TCDD-$^{13}C_{12}$ and 2,3,7,8-TCDF-$^{13}C_{12}$ to serve as internal standards for recovery calculations and quantification.

The HRGC/HRMS system was calibrated by comparing responses of known quantities of PCDD/PCDF isomers with a known quantity of 2,3,7,8-TCDD-$^{13}C_{12}$, 2,3,7,8-TCDF-$^{13}C_{12}$, and OCDD-$^{13}C_{12}$ internal standards. PCB analyses were done on solvent extracted XAD-2 solid sorbent by high-resolution gas chromatography/mass spectrometry.[25-27]

For Phase 1 of the fluidized bed evaluation, only TCDD and TCDF of the PCDDs and PCDFs and their 2,3,7,8 isomers were determined. In Phase 2, quantitative measurements were made of TCDD and TCDF in all effluent streams of the reactivator unit and penta-, hexa-, hepta-, and octa-chlorinated dioxins (CDDs) and furans (CDFs) in a selected number of samples. Additionally, samples of the natural gas used to fire the furnace were collected for PCB analysis. TCDD and TCDF concentrations were evaluated during virgin carbon reactivation with the infrared furnace. During a subsequent reactivation with this furnace, CDDs and CDFs were also determined.

Results

Fluidized Bed Reactivation

Samples were collected and analyzed for each sampling point identified in Table 13. In Phase 2 of the fluid bed evaluation, various tests were run to see: (1) if PCBs were in the natural gas going to the furnace or in the backup fuel oil supply, (2) the effect of 1-sec and 2-sec afterburner retention, and (3) the overall effect of the afterburner. Because of the more complex testing scheme, the test for this phase is shown in Table 13.

Phase 1 Results

In Phase 1, the carbon was heavily loaded with total organic halogens when chlorine disinfection was used prior to GAC adsorption. Reactivation of this carbon showed that the TOX concentration was reduced to approximately the instrument detection limit. Total organic halogen concentrations in the virgin, spent, and reactivated carbon were none detected (detection limit was 50 mg/kg), 657 mg/kg, and 55 mg/kg, respectively. Comparison between TOX concentrations on the virgin and reactivated carbon raises the question as to how much of the adsorbed TOX formed TCDD and TCDF during reactivation. It appears that only a small fractional amount of TCDD and TCDF were formed from the total TOX adsorbed on the carbon.

The 2,3,7,8-TCDD isomer was seen in the particulate stack emission discharges in low concentrations (0.001 to 0.02 ppt by volume). It was also seen occasionally in the scrubber water, probably attached to the particles removed in this process (0.06 to 0.07 ppt by weight). Relatively high concentrations of total TCDD were detected in the cyclone catch (as shown in Table 14). The temperature in the drying zone (149°C) was conducive to TCDD and TCDF formation and a large amount of GAC fines was available for attachment.

In two of the four tests, 2,3,7,8-TCDF was detected in the stack emission particulates (0.004 and 0.02 ppt by volume). However, in three tests, 2,3,7,8-TCDF was seen in the cyclone catch (0.6 to 2.7 ppb by weight). Total TCDF concentrations are shown in Table 15.

Previous studies had shown that PCBs could be present in Cincinnati natural gas.[28] Because PCBs can be precursors for furan formation, PCB analyses were done for the Cincinnati gas supply that was used to fire the fluid bed furnace. The average PCB concentration for Phase 1 was 228 ppt. Tri- and tetrachlorobiphenyls were the principal PCBs detected.

Table 13. Phase 2 Fluid Bed Reactivation, Cincinnati Water Works

Test	Reactivator Fuel[a]	Afterburner Off/On	Spent GAC	React. GAC	Scrubber Water	Cyclone Fines	AB[c] 1 sec	AB[c] 2 sec	Recup. Inlet	Fuel Stack[c]	Supply
1	NG	Off	X	X	X	T			TP		P
2	NG	Off	X	X	X	X			T		X
3	NG	On	X	X	T	X	T			T	X
4	NG	On	X	X	X	X		T			
5	NG	On	TP	TP	TP	X		TP		TP	
6	NG	Off	X	X	X	TP			TP		X
7	FO	Off			X				T		P

aNG: natural gas; FO: fuel oil.
bX: sample collected and archived.
T: sample collected and analyzed for TCDD/TCDF *only.*
TP: sample collected and analyzed for PCDD/PCDF.
P: sample collected and analyzed for PCB.
cModified Method 5 train with a sorbent (XAD-2) used for each test.

Table 14. TCDD Concentrations During GAC Reactivation at Cincinnati, Ohio

Reactivator Stream	Total TCDD			
	Test 1	Test 2	Test 3	Test 4
Spent carbon feed (ppb by wt)	NDa	ND	ND	ND
Cyclone catch (ppb by wt)	3.3	0.2	4.1	1.3
Scrubber water (ppt by wt)	ND	0.2	0.1	ND
Quench water (ppt by wt)	ND	ND	ND	ND
Stack emissions				
Particulates (ppt by vol)	0.02	0.02	0.004	0.01
Gas phase (ppt by vol)	ND	0.004	ND	ND

aND: TCDDs were not detected.

Phase 2 Results

In Phase 2, the TOX of the spent carbon, when chlorine was not added prior to GAC adsorption, was lower than when it was added prior to GAC adsorption (416 mg/kg vs 657 mg/kg). Subsequent reactivation produced a concentration of < 50 mg/kg, the same concentration as the virgin carbon. No 2,3,7,8-TCDDs were found. Essentially, no total TCDDs were detected except in the recuperator inlet when the afterburner was off. With the afterburner on, no CDD homologues below hepta-CDD were detected in the spent GAC, reactivated GAC, scrubber water, afterburner, or stack emissions. Concentrations of hepta-CDD and octa-CDD ranged from 0.001 ppt to 0.05 ppt and 0.006 ppt to 0.28 ppt, respectively.

Chlorinated dibenzofurans were detected in the afterburner during Phase 2 for all of the homologues evaluated (tetra- through octa-). This could possibly be due, in part, to the PCBs present in the natural gas supply that was used to fire the furnace. Only occasional small concentrations of PCDF were detected in the spent GAC and stack emissions.

The average PCB concentration in the natural gas supply was 90 ppt in Phase 2. Only chlorobiphenyl was detected.

Table 15. TCDF Concentrations During GAC Reactivation at Cincinnati, Ohio

Reactivator Stream	Total TCDF			
	Test 1	Test 2	Test 3	Test 4
Spent carbon feed (ppb by wt)	NDa	ND	ND	ND
Cyclone catch (ppb by wt)	0.5	3.3	2.1	0.8
Scrubber water (ppt by wt)	ND	ND	ND	ND
Quench water (ppt by wt)	ND	ND	ND	ND
Stack emissions				
Particulates (ppt by vol)	0.03	ND	0.03	0.02
Gas phase (ppt by vol)	0.01	0.006	ND	ND

aND: TCDFs were not detected.

Table 16. Effect of Chlorination on Total TCDD Concentration

	Prechlorine		No Prechlorine	
Reactivator Sample	Total TCDD[a]	TOX[b]	Total TCDD[c]	TOX[b]
Spent GAC feed (ppb by wt)	ND[d]	657	ND	416
Reactivated GAC (ppb by wt)	e	55	ND	<50
Cyclone catch (ppb by wt)	2.2	e	0.01	e
Scrubber water (ppt by wt)	0.1	e	e	e
Quench water (ppt by wt)	ND	e	e	e
Stack emissions				
Particulates (ppt by vol)	0.014	e	ND	e
Gas phase (ppt by vol)	0.001	e	< 0.001	e

[a]Average for four tests, ppt or ppb concentrations as indicated.
[b]TOX concentration in mg/kg.
[c]Average of two tests.
[d]ND: Not detected.
[e]Not sampled.

Effect of Chlorination

During Phase 1 and Phase 2, similar furnace reactivation conditions were maintained. Therefore, without the high-temperature afterburner in operation, data collected during reactivation of GAC that received chlorinated water (Phase 1) were compared with data from GAC that received unchlorinated water (Phase 2).

Comparable sampling points for the two phases were spent GAC, cyclone catch, and stack. Table 16 shows that no total TCDD was detected in the spent GAC for either phase, although the TOX concentration was higher when chlorine was added prior to GAC adsorption. This higher TOX is reflected in the total TCDD concentration of the cyclone catch and stack effluent. Total average TCDD concentrations were 2.2 ppb, 0.014 ppt, and 0.001 ppt for the cyclone catch, stack particulates, and stack gas, respectively, when chlorine was added prior to GAC. Conversely, with no chlorination prior to GAC, total average TCDD concentrations were 0.01 ppb, not detected, and < 0.001 ppt for the cyclone catch, stack particulates, and stack gas, respectively.

The same general trend occurred for the total TCDF where no chlorine prior to GAC produced lower concentrations than when chlorine was used. The exceptions noted were the spent GAC and stack gas (Table 17).

Afterburner Effect

Tables 18 and 19 show the effect of the high-temperature afterburner on the concentration of CDD and CDF homologues. With the afterburner off, no CDD homologue concentration was detected in the recuperator particulates below hepta-CDD. Except for tetra-CDD, concentrations were seen for all the other homologues (penta- through octa-) for the gas phase. With the afterburner on, no CDD homologues below hepta-CDD were present in

Table 17.   Effect of Chlorination on Total TCDF Concentration

Reactivator Sample	Prechlorine Total TCDF[a]	No Prechlorine Total TCDF[b]
Spent GAC feed (ppb by wt)	ND[c]	0.07
Reactivated GAC (ppb by wt)	d	ND
Cyclone catch (ppb by wt)	1.7	0.3
Scrubber water (ppt by wt)	ND	ND
Quench water (ppt by wt)	ND	d
Stack emissions		
Particulates (ppt by vol)	0.02	ND
Gas phase (ppt by vol)	0.004	0.052

[a]Average of four tests.
[b]Average of two tests.
[c]ND: not detected.
[d]Not sampled.

Table 18.   Effects of an Afterburner on CDD Concentration

Sample	Tetra-	Penta-	Hexa-	Hepta-	Octa-	Total
Afterburner off[a]						
Particulate (ppt by vol)	ND[b]	ND	ND	0.012	0.068	0.080
Gas phase (ppt by vol)	< 0.001	0.008	0.021	0.012	0.011	0.052
Afterburner on[c]						
Particulate (ppt by vol)	ND	ND	ND	0.014	0.048	0.062
Gas phase (ppt by vol)	ND	ND	ND	0.001	0.006	0.007

[a]Sample collected after recuperator.
[b]ND: not detected.
[c]Sample collected after 2-sec residence time.

Table 19.   Effects of an Afterburner on CDF Concentration

Sample	Tetra-	Penta-	Hexa-	Hepta-	Octa-	Total
Afterburner off[a]						
Particulate (ppt by vol)	ND[b]	< 0.001	ND	ND	0.019	0.019
Gas phase (ppt by vol)	0.103	0.073	0.046	0.042	0.018	0.282
Afterburner on[c]						
Particulate (ppt by vol)	ND	0.001	0.002	0.004	0.002	0.009
Gas phase (ppt by vol)	0.007	0.003	0.002	0.001	0.001	0.014

[b]ND: not detected.
[a]Sample collected after recuperator.
[c]Sample collected after 2-sec residence time.

either the particulate or gas phase. The total concentration of the CDD homologues (tetra- through octa-) after reactivation without an afterburner was 0.069 ppt.

For the particulates, only octa-CDF was detected with the afterburner off, while all other homologues except tetra-CDF were seen when the afterburner was on. In the gas phase, all of the CDF homologues were detected regardless of whether the afterburner was on or off. However, concentrations were lower when the afterburner was on. Total CDF concentration was 0.301 ppt with the afterburner off and 0.023 ppt with the afterburner on.

Infrared Reactivation

A series of tests was conducted during infrared reactivation of virgin GAC to determine if the carbon contained precursors that could produce TCDD or TCDF. No 2,3,7,8- or total TCDD or TCDF concentrations were detected in any stack effluent or process stream samples.

The TOX of the spent GAC that was exposed to chloraminated water at Jefferson Parish was 460 mg/kg. The reactivated GAC contained < 50 mg/kg TOX (detection limit). During infrared reactivation of the spent GAC at Jefferson Parish, three tests were run for PCDDs and PCDFs. As shown in Table 20, no CDD homologues below hexa-CDD were detected in the stack emissions, but 0.18 ppt TCDD was found in the scrubber water. Concentrations of hexa-CDD (ND–0.025 ppt), hepta-CDD (< 0.001 ppt–0.001 ppt), and octa-CDD (0.005 ppt–0.009 ppt) for the particulate and gaseous portions of the stack emissions were low. No tetra-CDD was found in the spent GAC, reactivated GAC, or drinking water samples. In one of the three tests, 0.16 ppt of TCDD was detected in the quench water.

All of the CDF homologues were detected in the gaseous portion of the stack emissions (Table 21). Below hexa-CDF, no homologues were detected in the particulate portion of the stack emissions. Tetra-, hepta-, and octa-CDF concentrations ranged from 0.03 ppt to 0.40 ppt in the scrubber water. Spent GAC contained an average of 0.17 ppb tetra-CDF, but none was found in the reactivated GAC, quench water, or drinking water samples.

Cancer Risk Assessment

Cancer risk assessments of PCDD and PCDF emissions from the pilot fluidized bed furnace and a proposed full-scale carbon reactivation furnace at the Cincinnati Water Works and the pilot infrared furnace at the Jefferson Parish Water Works were estimated to determine potential carcinogenic risk to the exposed population. For exposure analysis, the Human Exposure Model (HEM) was used to predict maximum annual average ground level concentration of 2,3,7,8-TCDD equivalence and produce quantitative

Table 20. PCDD Concentrations from Infrared Reactivation at Jefferson Parish, Louisiana

	2,3,7,8-TCDD	Total TCDD	Total Penta-CDD	Total Hexa-CDD	Total Hepta-CDD	Total Octa-CDD
Spent GAC (ppb by wt)	ND[a,b]	ND[b]				
Reactivated GAC (ppb by wt)	ND[b]	ND[b]				
Scrubber water (ppt by wt)	0.09[b]	0.18[b]	ND[c]	ND[c]	ND[c]	0.20[c]
Quench water (ppt by wt)	0.05[b]	0.05[b]				
Drinking water (ppt by wt)	ND[b]	ND[b]				
Stack emissions						
Particulates (ppt by vol)	ND[b]	ND[b]	ND[d]	ND[d]	<0.001[d]	0.005[d]
Gas phase (ppt by vol)	ND[b]	<0.001[b]	ND[b]	0.025[b]	0.001[b]	0.009[b]

aND: not detected.
bAverage of three tests.
cRun during one test only.
dAverage of two tests.

Table 21. PCDF Concentrations from Infrared Reactivation at Jefferson Parish, Louisiana

	2,3,7,8-TCDF	Total TCDF	Total Penta-CDF	Total Hexa-CDF	Total Hepta-CDF	Total Octa-CDF
Spent GAC (ppb by wt)	ND[a,b]	0.17[b]				
Reactivated GAC (ppb by wt)	ND[b]	ND[b]				
Scrubber water (ppt by wt)	0.03[b]	0.20[b]	ND[c]	ND[c]	0.03[c]	0.21[c]
Quench water (ppt by wt)	ND[b]	ND[b]	ND[d]			
Drinking water (ppt by wt)	ND[b]	ND[b]				
Stack emissions						
Particulates (ppt by vol)	ND[b]	ND[b]	ND[d]	<0.001[d]	<0.001[b]	0.001[d]
Gas phase (ppt by vol)	0.005[b]	0.014[b]	0.005[b]	0.002[b]	0.001[b]	0.002[b]

aND: not detected.
bAverage of three tests.
cRun during one test only.
dAverage of two tests.

Table 22.   Toxic Equivalency Factors Used in Estimating 2,3,7,8-TCDD Equivalents

Compound(s)	Relative Potency Factor (2,3,7,8-TCDD = 1)
2,3,7,8-TCDD	1.0
Other TCDDs	0.01
Penta-CDDs	0.5
Hexa-CDDs	0.04
Hepta-CDDs	0.001
Octa-CDDs	0.000
2,3,7,8-TCDF	0.1
Other TCDFs	0.001
Penta-CDFs	0.1
Hexa-CDFs	0.01
Hepta-CDFs	0.001
Octa-CDFs	0.000

expressions of public exposure to ambient air concentrations of pollutants emitted from the facilities. The HEM contains an atmospheric dispersion model with meteorological data and a population distribution estimate based on Bureau of Census data. Input data from the GAC reactivation furnace studies needed to operate the model included plant location, stack height and diameter, stack temperature, stack exit velocity of the offgases, and pollutant emission rate.

The concentration of the pollutant and the number of people estimated by the HEM to be exposed to PCDDs and PCDFs were identified for approximately 160 receptors (10 receptors along each of 16 wind directions located around the reactivation furnace). The pollutant concentrations and populations were paired to produce exposure estimates. Conservative assumptions (population is exposed continuously to the maximum annual average ground level concentration 24 hr/day for a 70-year lifetime) were used in the estimation of the probability of cancer occurring in the exposed population.

The cancer risk assessment methodology used a weighting system developed by EPA and based primarily on results from testing PCDDs/PCDFs in vitro through enzyme induction assays, cell keratinization assays, and in vivo information from long-term animal studies of 2,3,7,8-TCDD and hexa-CDD. The interim EPA procedure is a means of estimating the equivalent 2,3,7,8-TCDD–like activity of a complex mixture of CDDs and CDFs.[29,30] The weighting system consisted of 2,3,7,8-TCDD Toxic Equivalency Factors (TEFs) for each homologue of PCDD/PCDF that distinguished between the biologically active analogues of 2,3,7,8-TCDD within each homologous group from the non-2,3,7,8-substituted compounds. Multiplication of the homologue and 2,3,7,8-TCDD/TCDF concentrations measured in the stack gas by the TEFs in Table 22 converts the PCDD/PCDF mixture into 2,3,7,8-TCDD equivalents. The 2,3,7,8-TCDD equiva-

Table 23.   Risk Exposure from GAC Reactivation

Type	Reactivator Size (kg/hr)	Compounds	Afterburner	Maximum Individual Cancer Risk[a]
Fluidized bed	265[b]	TCDD	[c]	1.3×10^{-7}
Fluidized bed	265[b]	2,3,7,8-TCDD equivalence	off	5×10^{-8}
Fluidized bed	265[b]	2,3,7,8-TCDD equivalence	on	5×10^{-11}
Fluidized bed	1,141[d]	2,3,7,8-TCDD equivalence	off	4×10^{-8}
Fluidized bed	1,141[d]	2,3,7,8-TCDD equivalence	on	6×10^{-11}
Infrared	98[b]	2,3,7,8-TCDD equivalence	on	3×10^{-9}

Source: Lykins, Clark, and Cleverely.[32]
[a]Defined as the probability of cancer to an individual continuously exposed for 70 years to the maximum ambient concentration of 2,3,7,8-TCDD equivalence.
[b]Pilot plant.
[c]No afterburner in reactivation system.
[d]Estimated full-scale from pilot plant.

lent concentration can be factored with EPA's current estimate of carcinogenic potency for inhalation exposure to 2,3,7,8-TCDD.

The estimated upper limit of incremental excess cancer risk to the population exposed to PCDD/PCDF emissions from GAC reactivation at the Cincinnati and Jefferson Parish Water Works are summarized in Table 23. For the fluidized bed furnace at Cincinnati, cancer risks were determined for actual emissions sampled at the stack of a pilot plant and for a proposed full-scale reactivation system. In all cases, for the fluidized-bed and infrared furnace, the maximum individual excess cancer risks were lower than those reported for five municipal waste incinerators (U.S.EPA, 1981).[31] The highest risk value from these incinerators was 8×10^{-6} and it was determined that their discharge did not present a public health hazard for residents living in the immediate vicinity.

REFERENCES

1. DeMarco, J., et al. "Experiences in Operating a Full-Scale Granular Activated Carbon System with On-Site Reactivation," paper presented at 181st Meeting of American Chemical Society, Division of Environmental Chemistry, Atlanta, Georgia, March 30–April 3, 1981.
2. Juhola, A. J. "Relationships of Iodine Adsorption and Pore Structure of Activated Carbon," Carbon 13 (1975), pp. 437–442.
3. U.S. EPA-DWRD Contract 68-01-3487, "Determination of Dioxin Levels in Carbon Reactivation Process Effluent Streams," U.S. EPA Report 600/2-84-013 (March 1984).

4. "Interim Evaluation of Health Risks Associated with Emissions of Tetrachlorinated Dioxins from Municipal Waste Resource Recovery Facilities," U.S. EPA, Office of the Deputy Administrator (November 1981).
5. "Assessment of Emissions from a Recently Sampled Granular Activated Carbon Regeneration Facility," U.S. EPA, Office of Solid Waste and Emergency Response (December 1983).
6. Eiceman, G. A., R. E. Clement, and F. W. Karasek. "Analysis of Fly Ash from Municipal Incinerators for Trace Organic Compounds," *Analytical Chemistry*, 51(14):2343–2350 (1979).
7. Bumb, R. R., et al. "Trace Chemistries of Fire: A Source of Chlorinated Dioxins," *Science*, 210(4468):385–390 (1980).
8. Junk, G. A., and C. S. Ford. "A Review of Organic Emissions from Selected Combustion Processes," *Chemosphere*, 9 (1980), pp. 187–230.
9. Arthur D. Little, Inc. "Dioxin from Combustion Sources," American Society of Mechanical Engineers (1981).
10. "Dioxins," U.S. EPA Report 600/2-80-197 (1980).
11. Hutzinger, O., R. W. Frei, E. Merian, and F. Pocchiari, Eds. *Chlorinated Dioxins and Related Compounds*, Proceedings of Workshop held at the Instituto Superiore di Sanita, Rome, Italy (Elmsford, NY: Pergamon Press, 1980).
12. Tiernan, R. O., M. L. Taylor, J. G. Sloch, G. F. Vanress, and J. H. Garrett. "Characterization of Toxic Components in the Effluents from a Refuse-Fired Incinerator," *Resources and Conservation*, 9 (1982), pp. 343–354.
13. Browning-Ferris Industries, Inc. "Notes on Polychlorodibenzo Dioxins and Polychlorodibenzofurans in Connection with Waste-to-Energy Plants" (1983).
14. "Symposium Updates Health Effects of Dioxins, Benzofurans," *Chem. Eng. News* (September 1983), pp. 26–30.
15. Barnes, D. G. "Dioxins," *EPA Journal*, 9(3):16–19 (1983).
16. "Dioxin in the Environment: Its Effect on Human Health," American Council on Science and Health (1984).
17. Czuwa, J. M., and R. A. Hites. "Environmental Fate of Combustion-Generated Polychlorinated Dioxins and Furans," *Environ. Sci. Technol.* 16 (1984), pp. 444–450.
18. Rappe, C. "Analysis of Polychlorinated Dioxins and Furans," *Environ. Sci. Technol.* 18 (1984), pp. 78A–90A.
19. Godfrey, Jr., K. A. "Municipal Refuse: Is Burning Best?" *Civil Eng. ASCE* (April 1985), pp. 53–56.
20. Tschirley, F. H. "Dioxin," *Scientific American* 254(2):29–35 (1986).
21. DeMarco, J., and R. Miller. "Discovery and Elimination of Dioxins from a Carbon Reactivation Process" *J. Am. Water Works Assoc.* 80(3):66–73 (1988).
22. "Sampling and Analysis Protocol for Assessing Organic Emissions from Stationary Combustion Sources in Exposure Evaluation Combustion Studies." U.S. EPA Report 560/5-82-014 (1982).

23. G. R. Smithson, Jr. "Determination of Dioxin and Furan Levels in Effluent Streams of the Cincinnati Water Works Carbon Reactivation Systems," Quality Assurance Project Plan, Battelle, Columbus Laboratories, Columbus, OH (1984).

24. "Method 5 – Determination of Particulate Emissions from Stationary Sources," *Federal Register* (1977), p. 42–160.

25. "Feasibility Study of Granular Activated Carbon Adsorption and On-Site Regeneration," U.S. EPA Report 600/2–82–087A (1982).

26. "Determination of Dioxin Levels in Carbon Reactivation Process Effluent Streams," U.S. EPA Report 600/S2–84–013 (1984).

27. "Determining the Effectiveness of an Afterburner to Reduce Dioxins and Furans," U.S. EPA Report 600/S2–86–039 (1986).

28. "Sampling and Analysis of Dioxins and Furans During Granular Activated Carbon Regeneration Furnace Operation," Report to Jefferson Parish, Louisiana, Battelle, Columbus Laboratories, Columbus, OH (1986).

29. "Health Assessment Document for Polychlorinated Dibenzo-*p*-Dioxins." U.S. EPA Report 600/8–84–014F (1985).

30. "User's Manual for the Human Exposure Model (HEM)," U.S. EPA Report 450/5–86–001 (1986).

31. Barnes, D. G., and J. Bellini. "Procedure for Estimating Risks Associated with Exposures to Mixtures of Chlorinated Dibenzo-*p*-Dioxins and Dibenzofurans," presented at 5th International Symposium on Chlorinated Dioxins and Related Compounds, Bayreuth, West Germany (1986).

32. Lykins, Jr., B. W., R. M. Clark, and D. H. Cleverely. "Polychlorinated Dioxin and Furan Discharge During Carbon Reactivation," *J. Environ. Eng.* 114(2):300–316 (1988).

CHAPTER 8

Microbiology of GAC Filtration and Biological Activated Carbon

INTRODUCTION

It is well known that when granular activated carbon is used in treating drinking water, microorganisms develop on the filter granules and are found in the filtrate. European treatment practice has exploited this phenomenon by using a combination of ozone and carbon for removing organics from drinking water.[1]

In this process, often called biological activated carbon (BAC), it is suggested that through the preozonation of the water, some of the higher-molecular-weight humic substances can be oxidized into more biodegradable forms with lower molecular weight. These newly formed lower-molecular-weight organic compounds are potentially available as a food source for microbes on the GAC bed; therefore, preozonation may result in the availability of more adsorptive sites for the less biodegradable, more harmful organic compounds that are poorly oxidized. It is also possible that the newly formed lower-molecular-weight organics are more easily adsorbed. Treatment trains that incorporate biological activated carbon usually use the carbon process as the step before water is distributed. Recently, investigators have determined that microorganisms from BAC systems can colonize carbon fines that may enter the distribution system.

Treatment strategies designed to remove organics from raw source water must not create a pathway for the passage of pathogens into the finished water. Since there are a variety of treatment configurations using granular activated carbon that will remove varying amounts of raw water organics,

205

this section will evaluate available data from six of the field-scale operations and pilot-plant investigations of GAC treatment as it affects the microbial quality of process water, finished water, and drinking water in distribution systems.

GAC CHARACTERISTICS

A variety of carbon materials have been used to produce the adsorption surfaces essential to organic removal. Variations in adsorption capacity do occur among commercial supplies of GAC and are related largely to both carbon source material and the activation process employed by the manufacturer. Carbon sources include bituminous coal, bones, coconut shells, lignite, peat, pecan shells, petroleum-based residues, pulp mill black ash, and wood. Materials used in the full-scale and pilot-plant studies described herein were predominantly coal base or lignite (Table 1). While particle size of various granular activated carbons is fairly uniform (based on sieve sizing), surface areas for adsorption and resulting microbial attachment are related to source of GAC material. Lignite-derived GAC has approximately 60% of the total surface area of GAC produced from coal. Activation processes include treatment with an oxidizing gas (steam or air) at high temperatures and affect surface area. Other key characteristics of GAC include particle pore structure and total surface area. Void spaces between particles influence establishment of microbial colonization of the GAC filter bed. For this reason, the persistence and growth of organisms of public health concern (primary pathogens, opportunistic pathogens, and indicators) in GAC filters have drawn considerable attention in recent years.

Sanitary Quality of Virgin Carbon

In the various field-scale projects, startup operations revealed surprising initial coliform and heterotrophic bacterial contamination problems with GAC.[2] This contamination may have originated in undisinfected filter basins converted from sand filtration operations, GAC supplied by the manufacturer, or the water slurry movement of GAC into the filter compartment.

Full-scale evaluation of three different types of GAC placed into converted sand filter basins illustrates the possibility that these basins were not adequately cleaned prior to conversion. Initial influent quality for the first two months to new GAC beds brought no detectable total coliforms in 100-mL samples. These samples contained a free chlorine residual of 1.0 to 2.0 mg/L. However, filter effluent water from each of three different GAC materials (Filtrasorb 400, Filtrasorb C, and Hydrodarco 8 × 16) contained

Table 1.   Characteristics of Granular Activated Carbon

Manufacturer	Trade Name	Source Material	Sieve Size	Surface Area (m²/g)
Calgon	Filtrasorb C	Coal	12 × 40	—
Calgon	Filtrasorb 400	Coal	12 × 40	1050–1200
Carborundum	GAC 40	Coal	12 × 40	1000–1100
ICI	Hydrodarco 1030	Lignite	10 × 30	600
Westvaco	WVW 8 × 30	Coal	8 × 30	800
Westvaco	WVW 14 × 40	Coal	14 × 40	+940
Westvaco	WVG 12 × 40	Coal	12 × 40	1050–1100

46 to 85 total coliforms per 100 mL during the first week's run. On both Filtrasorb materials, total coliform densities peaked in three weeks, then declined to less than one coliform per 100 mL by Week 13 or 14. The heterotrophic bacterial population rose to density levels of 104 organisms/ mL within four to five weeks, then declined to below 102 organisms/mL by the 11th week for both Filtrasorb materials. Hydrodarco GAC prepared from lignite did not exhibit any increase of coliforms above the startup density of 46 total coliforms per 100 mL in the effluent and proceeded to decline to below detectable levels by Week 11. Microbial growth on the Hydrodarco material peaked at 104 organisms/mL during Week 7, then declined to below 102 organisms/mL by the 11th week, paralleling similar microbial activity on the other two GAC materials. These findings suggest that available nutrients and declining water temperatures were important factors in microbial persistence in the GAC contamination problem.

GAC Reactivation

The adsorptive capacity of GAC is eventually exhausted after a period of a few months to several years (depending on the type and concentration of organics in the process water), and eventually there is a need to either replace the GAC or reactivate it for reuse. Reactivation generally involves passing the spent carbon through a furnace heated to oxidize adsorbed materials in a controlled atmosphere. This reactivation process results in some loss of material and may cause some particle size reduction and change in surface characteristics. What effect these subtle changes in physical characteristics have on bacterial adhesion to GAC and on microbial colonization of these reactivated particles is not known. Some insight into this aspect was obtained from a pilot-plant evaluation of virgin and reactivated carbon at Cincinnati and a similar study at a full-scale site at Manchester, New Hampshire, where the GAC was cycled through three periods of use and reactivation.

Product water from the sand filters in the Cincinnati water treatment train was fed to parallel 0.9-m (2.6-ft) columns of either virgin carbon (WVG 12 × 40) or reactivated GAC of the same origin. Effluent turbidity measurements from both columns during the first 63 days were often higher than found in the influent water (sand filter effluent). Because turbidity measurements do not detect carbon fines, these increased values represented intermittent passage of entrapped influent turbidity particles during initial stabilization of the packed column.

More significant releases of heterotrophic bacteria occurred in the virgin GAC once the water temperature began to rise to over 10°C than were noted with the use of reactivated GAC. During this warm water period, chlorine residuals were generally undetectable in these effluents. There were only infrequent coliform occurrences recorded in either column effluent during this period; one coliform per 100 mL on two occasions in virgin GAC, and 1, 7, and 12 coliforms per 100 mL in three effluents from the reactivated carbon. Only on one day were there concurrent coliform occurrences in effluents from both carbon columns. No coliforms per 100 mL were detected in any influent samples. Correlation of coliform occurrences with increased turbidity releases was inconclusive because of the limited data available based on one sample per week.

The field study in Manchester, New Hampshire using GAC filters provided evidence that microbial activity on virgin GAC (Westvaco WVW 8 × 30) and the same GAC material reactivated were identical. While water temperature above 10°C does encourage coliform regrowth, the essential element is in the bacterial quality of processed water applied to these filters. Data from an initial run indicated the microbial quality remained excellent during 112 days of operation. This occurred 41.2% of the time in the spring with water temperatures above 10°C. However, during the second run that started in early autumn, a different pattern of bacterial quality of the filter effluents was noted. GAC influent water from the sand filter was observed to contain 1 to 8 total coliforms per 100 mL in a two-week period during September 1980. With warm water temperatures, some colonization of coliforms occurred and there was passage of these organisms through both GAC filters intermittently during a 7- to 11-week period before declining to nondetectable levels per 100 mL. Heterotrophic bacterial densities rose two weeks before coliforms were detected and declined over the same time period.

Turbidity measurements did not correlate with these bacterial quality changes in the GAC filter effluents. The succeeding run of virgin GAC in parallel with GAC reactivated a third time produced the same high-quality effluents noted in the first run. Again, the influent GAC process water was excellent. These data demonstrate that if influent water is of excellent bacteriological quality, transient microbial persistence with only minimal regrowth will result whenever some passage of organisms occurs in the pretreatment stages. Passing this high-quality process influent water into either the virgin GAC or reactivated GAC filters produced similar bacteriological qualities in the filter effluents.

Pressure Contactors

While granular activated carbon is often used in conventional treatment beds designed originally for sand filtration, GAC may also be used in pressure contactors. In this latter configuration, GAC bed depth is usually more than 0.9 m (36 in.), the normal rapid filter depth, to provide sufficient contact time necessary to remove certain classes of organics as the water is pumped through each unit.

Since little information is available in the published literature on the bacteriological aspects of pressure contactors, measurements for heterotrophic bacteria and total coliforms in the effluent from these devices were included in various pilot and field-scale operations. Factors to be considered were the extent of bacterial colonization in fast-flowing water through a deeper GAC bed and containment in a closed cylinder. Three different virgin GAC materials (Hydrodarco 10 × 30, WVW 12 × 40, and Filtrasorb 400) were evaluated simultaneously in a pilot plant. Influent water to the contactors was finished water containing 0.6 to 1.3 mg/L free chlorine residual.

Influent data indicated the finished water to be of high quality throughout the study period. One weakness with this interpretation was the sampling interval which was daily for the first four days and then once every two to four days thereafter. Such infrequent sampling does not recognize intermittent pulsing of low-density levels of microorganisms into the pressure contactors. Coliforms were detected in the effluent from all contactors during Days 1 to 4 and from the Filtrasorb 400 contactor on Days 24 to 63. Colonization and subsequent breakthrough was unlikely because of cold water temperatures.

A study of the heterotrophic bacterial population in different systems in this report suggests some differences in density magnitudes that does not relate to water temperature, type of GAC, or process utilized. These inconsistencies are a reflection on the Standard Plate Count protocol which recently has been demonstrated to detect less of the heterotrophic bacterial population than R2A medium, and on use by some investigators of extended incubation time (seven days) and 28°C incubation temperature rather than 35°C.

Use of reactivated GAC (Hydrodarco 10 × 30) in contactors with GAC bed depth of 2.0 m (6.5 ft) was investigated in one phase of another pilot study at Evansville, Indiana. In this situation, the influent water was of lesser quality than the chlorinated finished water used in other studies such as Passaic. Ohio River water was predisinfected with chlorine dioxide, chemicals were added, and the water was flocculated, settled, and passed through a mixed-media filter to become influent water for the GAC contactors.

Virgin Hydrodarco 10 × 30 GAC was used for two months with this pilot-plant water system. During this test period, heterotrophic bacterial densities increased to 3 logs of the influent within four days, and one coliform breakthrough occurred on Day 12, when 64 total coliforms and 53 total coliforms per 100 mL were detected in the influent and effluent, respectively. No coliforms were detected before or after that day, suggesting colonization was not a factor.

At termination of this 60-day run, the treatment process was discontinued, and the GAC was removed from the contactor and shipped to an offsite location for reactivation. Upon return, this reactivated GAC was placed in the contactor and the testing restarted. On restarting the pilot plant, influent water quality for the first seven days was below expectations; turbidity ranged from 1.3 to 24.0 NTU and total coliform densities varied from 7 to 36 organisms per 100 mL. As anticipated, coliforms occurred in the GAC effluent and varied in densities from 1 to 8 organisms per 100 mL for these same seven days. After that interval, the total coliform occurrence in the contactor effluent was limited to Day 28, when one total coliform per 100 mL was found. Coliform passage in elevated turbidities during the period of bed stabilization was probably the mechanism for migration. Coliform colonization on the reactivated GAC was unlikely, because the low density for these bacteria never increased above coliform counts per 100 mL observed in the influent water.

Heterotrophic bacterial densities in the influent to the GAC contactor were also initially high, but declined in reduced cyclic fashion to reach densities consistently below 10 organisms/mL by Day 36 to the end of the experimental run on Day 63. This trend was also detected in data on heterotrophic bacterial densities obtained from the contactor effluent. The initial run with virgin GAC and the second run after the GAC was reactivated did have measurable chlorine dioxide residuals in the influent water which were completely adsorbed during passage through the contactor. Some interesting observations were noted in data collected from a pilot-plant investigation in which four pilot contactors were placed in series and fed with effluent from the sand filters of the water plant at Jefferson Parish.

Data reported for the first 30 days of operation indicated influent water had no detectable coliforms per 100 mL and heterotrophic bacterial densities were below 50 organisms/mL. As the influent water passed through the series of contactors, there was a progressive increase in low-density coliform occurrences. These results suggest passage of coliforms in densities below 1 organism per 100 mL from the influent with no subsequent establishment of significant coliform colonization in any of the contactors. Sequential data indicated coliforms (1 organism per 100 mL) were detected in the individual columns in series on Days 138, 118, 40, and 40, respectively.

This pattern was not characteristic of the heterotrophic bacteria, which were quick to colonize within three days and release organisms that were 3 logs higher in densities than detected in the influent. The heterotrophic bacterial colonization did not persist even in the warm water temperatures, declining to below 200 organisms/mL by Day 107 for all columns. Apparently, available nutrients eventually became the limiting factor.

GAC Process Placement in Water Treatment Train

Optimizing removal of organics in process water, in theory, would suggest the GAC process should be placed early in the progressive treatment of polluted surface waters. In fact, powdered carbon is often applied in water impoundments to control taste and odor problems, but it is not adequate for removal of many other organics. Obviously, the GAC filtration process can not be applied to raw surface waters with significant turbidities (above 1 NTU) because silt in these waters quickly coats the carbon particles and rapidly reduces organic adsorption capacity. Thus, settling of raw water and chemical treatment with clarification, at least, should proceed the GAC process. These preceding treatment processes also remove much of the turbidity associated with microbial flora (a wide range of environmental organisms, some of which are capable of aggressive colonization on GAC particles). What impact placement of the GAC process following clarification, or after sand filtration and with modifications in type of disinfectant or point of application, has on microbial quality of the effluent was studied in a variety of field experiences.

Placement of a GAC filter adsorber on-line after chemical clarification did result in an approximate 85% reduction in turbidity to NTU values ranging from 0.3 to 0.9 at Jefferson Parish. The occasional wide differences in residual turbidities reflect the entrapment of coagulant particles on the GAC bed and their migration through the filter prior to backwashing. Application of chloramines to the clarified water prior to passage through the GAC filter adsorber did not result in a complete reduction of total coliforms below the 100-mL baseline, except for the autumn period. Disinfectant concentration and contact time become more critical when chloramines are applied, since these agents are slower acting than free chlorine. Also, total chlorine residual data include not only the active disinfectant components, but some complexes of no disinfection power. As a consequence, coliforms were often found in the GAC filter adsorber effluent except during the winter periods. This period of absence more likely reflected only a brief decline in filter adsorber colonization, unlike the prior winter startup period when some breakthrough occurred during filter adsorber stabilization.

Moving the GAC adsorber treatment process to after sand filtration did result in an improvement (0.1 to 0.3 NTU) in effluent turbidity. The most beneficial effect was the improvement in the bacteriological quality of the influent. Heterotrophic bacterial densities were below 75 organisms/mL and no total coliforms were found in any of the effluent samples. This water quality improvement was a result of sand filtration effectiveness and increased contact time with chloramines added after clarification. As a result of better quality influent water to the GAC adsorber, no coliforms were detected in the GAC effluent over a three-year study period. However, little difference was observed in the cyclic rise and decline of the heterotrophic bacterial population (in filter adsorber and adsorber effluents) associated with either influents.

Prechlorination of the raw source water premixed with chemicals, flocculated, ammoniated, and then clarified before passage through a filter adsorber was also investigated. Chloramine residuals in the pilot plant influent water to the GAC filter adsorber ranged from 1.9 to 3.7 mg/L. No total coliforms were detected; the heterotrophic bacterial densities were between 2 and 17 organisms/mL. After passage through the GAC, no coliforms were detected, but the heterotrophic bacterial densities were 2 to 3 logs higher for the entire 73 days of operation. Water turbidity was reduced to approximately 8 to 23% of the influent values.

Parallel data were also obtained for influent water receiving the above treatment plus sand filtration. Again, no coliforms were detected and the heterotrophic bacteria rapidly colonized the GAC adsorber to densities similar to those found for the GAC filter adsorber. Effluent turbidity reductions for the adsorber and filter adsorber were similar. As expected, the adsorber was more effective in removing TOC over 65 days than the filter adsorber, probably a reflection of GAC particle coating by coagulant residuals from the clarifier effluent.

Shifting disinfectant application to only breakpoint chlorination of the clarifier effluent was also investigated at Thornton, Colorado. While total coliforms were not found, the influent heterotrophic bacterial population was observed to be somewhat higher. With the additional chlorine, contact time, and passage through the sand filter, bacterial quality of the influent process water to the GAC adsorber was excellent.

Applying chlorination only to the raw water to achieve breakpoint was studied in another mode with the filter adsorber and adsorber pilot columns. As anticipated, there was some fluctuation in heterotrophic bacterial densities and one coliform occurrence in the clarifier effluent, but sand filtration produced a high-quality effluent to the GAC columns. In both configurations, the end results were similar: no total coliforms were found, the heterotrophic bacteria rapidly colonized the GAC, and the bacterial counts proceeded in a progressive downward cyclic pattern towards densities below 1000 organisms/mL by Day 133 for the adsorber effluents and Day 140 for the filter adsorber. Again, TOC measurements did not correlate with microbial activity found in the GAC effluents.

BIOLOGICAL ACTIVATED CARBON

In order to assess the usefulness of biological activated carbon, a study was conducted by the city of Philadelphia to examine whether ozonation used as a pretreatment to GAC adsorption can increase the useful bed life of GAC sufficiently to justify its cost.[3] The project was based on the understanding that ozonation will transform some of the higher-molecular-weight humic substances into more readily biodegradable forms. These lower-molecular-weight organic compounds are then potentially available as a food source for microbes already present on the GAC bed. Preozonation may thus make available more adsorptive sites on the carbon for the less biodegradable, more harmful organic compounds that are poorly oxidized (e.g., chloroform and 1,2-dichloropropane). Preozonation may also help prolong GAC life by stripping volatile organic compounds from the process stream.

The project was conducted on both a pilot and laboratory scale to investigate the feasibility of incorporating an ozonation and/or a GAC unit process into a conventional water treatment train to remove trace organics of health concern. The relationship between adsorption and biological activity during water treatment with ozone and GAC was carefully evaluated, as were the effects of prechlorination on these mechanisms. The removal of trace organic substances of health concern at the nanogram- to microgram-per-liter level and the removal of TOC at the milligram-per-liter level were monitored along with microbial parameters of biological speciation and growth rates.

The following questions were addressed:

1. Can enhanced TOC removal before GAC treatment increase the capacity of the GAC for the trace organics of health concern?
2. Can ozonation as a pretreatment to GAC adsorption increase the useful life of GAC sufficiently to justify its cost?
3. How does prechlorination affect the ozonation process and the GAC adsorption capacity for chlorinated organics and other volatile organics?
4. Can a conventional treatment system be maintained without predisinfection? Can the bacterial integrity of the distribution system be maintained with only postchlorination?

Three major areas of study were investigated to understand the use of both ozone and GAC for water treatment. These were:

1. laboratory-scale studies of the ozonation process
2. pilot-scale studies of the ozone/GAC process
3. mini-column studies that examined mechanisms for long-term TOC removal

Ozonation Studies—Laboratory

The overall objective of the laboratory ozone study was to develop an understanding of the ozone process for treatment of drinking water. This included development of a laboratory protocol that could be used to observe the effect of ozone treatment prior to larger-scale operation. The specific ozone studies were broken down into the following areas:

1. ozonation of Delaware River water at various stages of treatment; also, ozonation of high-molecular-weight materials (humics) which represent about 90% of the total organic carbon present in the river water
2. investigation of ozone utilized for assessing ozone demand
3. ozonation kinetics of different substrates present in the water

Pilot-Plant Studies

Five advanced water treatment (AWT) systems were evaluated in this study to determine their relative effectiveness for removing trace organic compounds from conventionally treated Delaware River water. Four of the five systems tested were GAC systems: one preceded by ozonation (O_3/GAC), another by chlorination (Cl_2/GAC), a third by both (Cl_2/O_3/GAC), and a fourth by neither (GAC). The fifth system was a sand bed preceded by ozonation (O_3/sand).

Conventional Treatment Train

Of the five AWT systems tested, two received their water from the chlorinated rapid sand filter effluent of the Torresdale Water Treatment Plant, a conventional coagulation/filtration plant which supplies the city of Philadelphia with one-half of its daily water requirements. The remaining three AWT systems received nonchlorinated rapid sand filter effluent of a 30,000-gpd pilot plant. The pilot plant was designed and built in 1976 to offer a high degree of flexibility for investigations of new and conventional water treatment techniques. Table 2 compares the operating characteristics of the conventional treatment trains of the Torresdale and pilot plants during the period under investigation. Possibly the most important of the many similarities between the two plants are that they draw their influent water from the same point along the Delaware Estuary, and that they both coagulate, settle, and filter their water. The primary difference between the operations of the plants is the fact that only the Torresdale Plant chlorinates its process water; the pilot plant does not practice disinfection in any manner. The purpose of using the two plants was to determine the effect of chlorination and chlorine by-products on the AWT systems investigated.

Table 2. Conventional Treatment Plant Comparisons

	Torresdale Plant	Pilot Plant
Average flow	230 mgd	30,000 gpd
Source	Delaware River	Delaware River
Intake frequency	Tidal cycle—every 6 hr	Continuous
Raw water basin	18 hr detention time	12 hr detention time
Coagulant	Ferric chloride	Ferric chloride
pH adjustment	Lime	Sodium hydroxide
pH	8.5	8.5
Clarification	Horizontal flow-through paddle flocculators and sedimentation basins	Upflow sludge contactors
Filtration	94 RSFs[a]—1041 ft2 each	2 RSFs—5 ft2 each
Chlorination	Breakpoint at rapid mix; 1.5 to 2.0 ppm free Cl_2 residual applied at RSFs	None
ClO_2 and $KMnO_4$	Occasional (at plant intake)	None
PAC	Occasional (at rapid mix)	None
Aeration	None	Occasional (at rapid mix)
Ammoniation and fluoridation	Posttreatment[b]	None

[a]Rapid sand filtration.
[b]Torresdale water to be applied to the AWT systems is drawn from the RSF effluents, prior to ammonia and fluoride addition.

Pilot Plant Description

The pilot plant drew its water from the Delaware River, just a few feet away from the intake gates of the Torresdale Water Treatment Plant. A submersible pump was used to continuously pump approximately 56,700 gpd from the river to a raw water sedimentation basin located 3000 ft from the intake. To protect the pump from any floating or suspended objects in the river, a 2- × 2- × 8-ft shaft was used. The shaft was constructed of cast iron sheeting and coarse stainless steel screening.

The raw water basin (RWB) had a capacity of 14,000 gal, allowing for a detention time of 12 hr at the design flow rate of 56,700 gpd. An important feature of the RWB is that it was covered with a wood roof, which prevented excessive algae growth in the undisinfected water. Additionally, an overflow piping system was used to relieve the basin of excess flow from the river pump. To protect the pilot plant system from unexpected river pump breakdowns, the RWB was also equipped with an electronic level-sensing device.

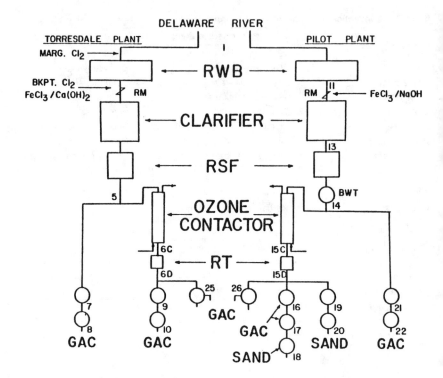

Figure 1.   Flow schematic of pilot-scale studies.

Advanced Water Treatment Systems

Following conventional treatment, the rapid sand filter effluents of the Torresdale and pilot plants were applied to the ozone and carbon contactors. Figure 1 is a schematic of the two conventional treatment trains and their respective AWT systems. The numbering system outlined in Figure 1 represents sampling sites. Table 3 describes those sampling sites relevant to the results to be presented in this chapter.

Ozonation Systems

Ozone gas was generated onsite with two ozone generators. Each unit was capable of generating 40 g/hr (2 lbs/day) of ozone from dry air. Typically, however, the generators were operated well below capacity, at a flow rate of 2 scfm and an ozone concentration of only 1.0–1.5 mg/L. As an oxygen source for the ozone generators, ambient air was drawn through two compressors and dried to a dew point of –40°F with heatless regenerative driers.

Stainless steel piping and Teflon tubing directed the ozone gas flow from the generators to the contactors. The contactors were constructed of 316L

Table 3. Sampling Locations for Pilot Study

Site Numbers		Site Description	Mnemonic
TPa	PPb		
1	11	Delaware River water	RIVER
4	13	Clarifier effluent	CLAR
5	14	Rapid sand filter effluent	RSF
6C	15C	Ozone contactor effluent	O_3C
6D	15D	Ozone retention tank effluent	O_3RT
7	21	GAC (15-min EBCT)	GAC_{15}
8	22	GAC (30-min EBCT)	GAC_{30}
9	16	Ozone/GAC (15-min EBCT)	O_3/GAC_{15}
10	17	Ozone/GAC (30-min EBCT)	O_3/GAC_{30}
—	19	Ozone/sand (15-min EBCT)	$O_3/sand_{15}$
—	20	Ozone/sand (30-min EBCT)	$O_3/sand_{30}$
25A	26A	Ozone/GAC (multi-tap) (7.5-min EBCT)	$MT_{7.5}$
25B	26B	Ozone/GAC (multi-tap) (15-min EBCT)	MT_{15}

aTP: Torresdale Plant.
bPP: Pilot Plant.

stainless steel pipe measuring 4 m (13 ft) in height and with an inside diameter of 25 cm (10 in.). Table 4 presents typical operating characteristics for the two ozone contacting systems.

Both ozone systems utilized co-current flow, where both the ozone gas and water enter the contactors from the bottom. For the diffusion of the ozone gas into water, a porous dome diffuser was used. These diffusers

Table 4. Ozonation System Operating Parameters

	Torresdale Plant	Pilot Plant
Contactor		
Type	Co-current	Same
Overall	4 m × 25 cm (13′ × 10″) diameter	Same
Material	Type 316L S.S.	Same
Diffuser	18 cm (7 in.) diameter (ceramic glass)	Same
Contact time (min)		
O_3 Contactor	22	Same
O_3 Retention Tank	25	Same
Flow rates		
Gas (scfh)	65	77
Water (gpm)	2.0	2.0
G:W ratio	4.1	4.8
Average ozone concentrations		
Applied dose (mg/L)	1.2	1.2
O_3 residual in water (mg/L)	0.3	0.3
O_3RT residual in water (mg/L)	0.0	0.0
Ozone utilizeda (mg/L)	1.2	1.6
Ozone transfer efficiency (%)	25	17

aOzone utilized $= (G:W)\,([O3]G_{in} - [O3]G_{out} - [O3]W_{out})$.

measured 18 cm (7 in.) in diameter and were made of ceramically bonded resins of aluminum oxide. Effective bubble size for the diffusers was rated at 0.2 mm. A polyvinyl chloride (PVC) bolt was employed as both the gas inlet and the diffuser fastener. Unfortunately, these bolts sheared in half and were therefore not properly securing the diffusers to the bottom support plates of the contactors. Fortunately, the identical situation was observed for both contactors, which allowed comparison of the two systems. The resulting low ozone transfer efficiency was countered through long ozone/water contact times and high gas:water ratios. Thus the desired oxidation reactions were achieved at higher operating costs than would normally be necessary.

Following ozonation, the water was passed through stainless steel retention tanks, allowing sufficient contact time to decrease the ozone residual in the water to below detectable limits. This minimized the ozone/water contact time and the oxidation of the carbon granules. Centrifugal pumps transferred the water from the retention tanks to the O_3/GAC and the O_3/sand systems.

Adsorption Systems

Listed below are the four adsorption systems investigated in this study and a brief description of each:

- GAC(TP): Chlorinated rapid sand filter effluent of the Torresdale plant (Site 5) is applied to two GAC columns in series (Sites 7 and 8, respectively). EBCT = 15 min each.
- O_3/GAC(TP): Chlorinated rapid sand filter effluent of the Torresdale plant (Site 5) is ozonated (site 6D) and applied to two GAC columns in series (Sites 9 and 10, respectively). EBCT = 15 min each.
- GAC(PP): Nonchlorinated rapid sand filter effluent of the pilot plant (Site 14) is applied to two GAC columns in series (Sites 21 and 22, respectively). EBCT = 15 min each.
- O_3/GAC(PP): Nonchlorinated rapid sand filter effluent of the pilot plant (Site 14) is ozonated (Site 15D) and applied to two GAC columns in series (Sites 16 and 17, respectively). EBCT = 15 min each.

Table 5 lists the design and flow characteristics of the GAC and O_3/GAC systems. Each system consisted of two columns operated in series. Each column was 1.8 m (6 ft) in length with a 15-cm (6-in.) inside diameter. The columns were connected with Teflon and stainless steel tubing. Each column had a 1.2-m (4-ft) carbon bed depth and an EBCT of 15 min.

All of the systems were operated under a pressure of approximately 15 psi. This allowed for the headloss to build up to 3–5 m (10–15 ft) of water before backwashing was initiated. The backwashing process, which was operated manually, was performed on all columns at the same time.

Table 5. Pilot Column Design

	GAC	O₃/GAC	MT	O₃/Sand
Construction	Glass	Glass	Teflon	Glass
Diameter, cm (in.)	15 (6)	15 (6)	4 (1.6)	15 (6)
Bed depth, m (ft)	1.2, 2.4	1.2, 2.4	0.5, 0.9	1.2, 1.4
	(4, 8)	(4, 8)	(1.5, 3)	(4, 8)
Media	Carbon[a]	Carbon[a]	Carbon[a]	Sand
Preozonation	No	Yes	Yes	Yes
SLR, L/min/m² (gpm/ft²)	81 (2)	81 (2)	61 (1.5)	81 (2)
EBCT, min	15, 30	15, 30	7.5, 15	15, 30

[a]Calgon F400 granular activated carbon.

All of the adsorption systems began operation at the same time. The study continued for 12 months, at which time all of the columns were removed from service with the exception of the $GAC_{15}(GAC)$ and $O_3/GAC_{15}(PP)$ systems, which remained in service for an additional 6.5 months.

In addition to the adsorption systems listed above, two "mini-scale" multi-tap systems were also operated in the O_3/GAC mode. A description of these two systems is also given in Table 5. The columns were constructed of 4-cm (1.6-in.)-diameter Teflon segments approximately 13 cm (5 in.) in length; each had a sampling port that allowed for the collection of both water and GAC samples. These systems were used to determine bacterial concentrations at various depths throughout the bed.

O_3/Sand System

In addition to the adsorption systems, an O_3/sand system was also investigated. This system was operated in parallel to the $O_3/GAC(PP)$ system. A description of this system may be summarized as:

O_3/sand(PP): Nonchlorinated rapid sand filter effluent of the pilot plant (Site 14) was ozonated (Site 15D) and applied to two sand columns in series (Sites 19 and 20, respectively). EBCT = 15 min each.

The design and flow characteristics of the O_3/sand system are listed in Table 5. The only difference between this system and the $O_3/GAC(PP)$ system was the sand medium in the O_3/sand system. Any organic removal observed through the O_3/sand system was due either to oxidation, volatilization, or biodegradation, since no adsorption took place in the O_3/sand system. Knowing the extent of each of the above then became a useful tool in determining the relative importance of biodegradation and adsorption in the carbon beds of the $O_3/GAC(PP)$ system.

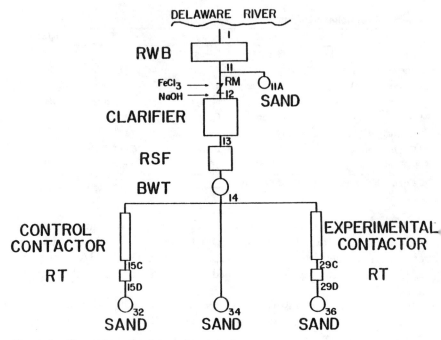

Figure 2.   Flow schematic of ozone/sand studies.

O₃/sand studies.   A six-month investigation of the O_3/sand system was initiated one year following the startup of the O_3/GAC pilot studies. The primary purpose of the O_3/sand investigation was to determine whether any significant increases in organic mineralization and/or biodegradability could be achieved through changes in the operating parameters of the pilot plant ozone contactor. Five studies were conducted. The operating parameters investigated were applied ozone dosage, ozone/water contact time, and direction of water flow through the contactor.

A general flow schematic for the five studies is shown in Figure 2. A description of the sites is listed in Table 6. In each study, one O_3/sand system was used for control purposes (the Control System), while the other system underwent process changes (the Experimental System). The

Table 6.   Sampling Locations for O_3/Sand Studies

Site Numbers			
Control System	Experimental System	Site Description	Mnemonic
14	14	Rapid sand filter effluent	RSF
15C	29C	Ozone contactor effluent	O³C
15D	29D	Ozone retention tank effluent	O_3RT
32	36	Ozone/sand (15-min EBCT)	$O_3/sand_{15}$
34		Sand (15-min EBCT)	$Sand_{15}$

unchlorinated effluent of the pilot plant rapid sand filter (Site 14) served as the influent to both ozone contactors. Following ozonation, the ozone retention tank effluents (Sites 15D and 29D) were applied to sand beds (Sites 32 and 36). Each sand bed had an EBCT of 15 min. This contact time was found sufficient for the bacteria on the sand beds to remove any biodegradable organics from the process stream. An additional sand bed (Site 34) received nonozonated pilot plant rapid sand filter effluent (Site 14). It was operated as a control system in order to determine whether any biodegradable organics were in the water applied to the ozone contactors. At each sampling location, samples were collected for DOC, UV 254, and THMFP analysis. In addition, both the halogenated and nonhalogenated volatile organic compounds measured during the pilot study were also tested.

A summary of the operating conditions for the O_3/sand studies is shown in Table 7. The objective of the first O_3/sand study, the ozone contactor performance comparison, was to determine whether the two ozone systems would yield equivalent results when operated under identical conditions. This study was crucial not only as a prelude to the other O_3/sand studies, but also as a check on the comparability of the results obtained from the two contactors during the O_3/GAC run.

The objective of the second study was to determine the influence of ozone/water contact time on the reduction of organic compounds from the process stream. This was accomplished by doubling the contact time in the experimental contactor.

In the third study, both the ozone dosage and the ozone/water contact time in the experimental system were doubled. Because of relatively high levels of DOC removal observed during this investigation, a fourth study was undertaken to look at the effects of increased ozone dose alone.

The objectives of the last study were to determine whether the relatively poor ozone transfer efficiency of the ozone contactor could be improved through a redirection of the water flow through the contactor. The control system was kept as a co-current flow contactor, while the experimental system was converted to a countercurrent system.

Mini-Columns

The overall purpose of this mini-column study was to determine the mechanisms of long-term TOC removal in a GAC contactor. An understanding of the mechanisms of organic removal in an adsorptive system is essential in developing predictive mathematical models and optimizing the adsorptive unit process for potable water treatment.

The overall purpose was accomplished through a series of experiments designed to separate the effects of adsorption and biodegradation in a GAC contactor. The major points addressed in the experiments were:

Table 7.   Operating Conditions for O_3/Sand Studies

	Control Ozonator			Experimental Ozonator		
Study	O_3 Dosage (mg/L)	Contact Time (min)	G:W[a]	O_3 Dosage (mg/L)	Contact Time (min)	G:W[a]
1. Ozone contactor performance comparison	1.2	22	4.6	1.2	22	4.4
2. Contact time comparison	1.5	22	4.6	1.5	44	9.2
3. Contact time and ozone dosage comparison	2.1	22	4.3	4.1	44	8.6
4. Ozone dosage comparison	2.2	22	4.6	4.0	22	4.1
5. Co-current vs countercurrent flow comparison	2.0	22	4.7	2.0	22	4.7

[a]G:W—gas to water ratio.

1. the relative contributions of bacterial degradation and slow adsorption to TOC removal in a GAC contactor
2. the reproducibility of column adsorption studies for inorganic, organic, and microbial parameters
3. the effect of temperature on the biological degradation and activated carbon adsorption from dilute concentrations of organics
4. the use of sand as an alternate medium for estimation of bioactivity in GAC contactors

To accomplish the major goals of the study, an experimental approach was required to differentiate between adsorptive and biological TOC removal. The experimental approaches used in this study were:

1. the development of an experimental method to inhibit biological TOC removal on GAC without affecting adsorption
2. the use of dissolved oxygen removal as an indirect measure of bioactivity in a GAC contactor

The mini-column work was conducted using effluent from the pilot plant ozone retention tank at the Torresdale Water Treatment Plant under field conditions (i.e., low and fluctuating organic concentrations with seasonal temperature variations). The parameter TOC was monitored for the following reasons:

1. Quasi–steady state removal had been demonstrated for this parameter at Philadelphia and elsewhere, suggesting biodegradation as the mechanism of removal.
2. Competition between specific organics and natural background organics as measured by TOC or humic substances had been demonstrated.
3. Humic materials act as precursors in the formation of THMs in potable water treatment. Humic materials comprise up to 90% of the natural background organic load found in natural waters. THMs and TOC have been correlated in potable water, although this has not yet been demonstrated in a general fashion.

Several small-scale experiments were conducted with the common focus of determining the mechanism of long-term TOC removal and quantifying the contribution of bacteria to the long-term removal.

Results

Conventional Treatment

Significant differences were observed in the DOC and THMFP effluent levels from the two conventional treatment systems. In fact, by postponing chlorination until after rapid sand filtration, the pilot plant conventional treatment processes reduced THMFP by one-third during the one-year period of study.

Two factors appear to have been primary influences. First, the two clarification systems varied greatly in their ability to remove DOC from their process streams. For example, the pilot plant clarifier (an upflow sludge contactor without clarification) removed nearly three times the DOC removed through the Torresdale clarification system (transverse flow-through paddle flocculators and sedimentation basins with chlorination). Second, a portion of the DOC normally found in the Delaware River Estuary was readily degraded by the bacteria present in the pilot plant conventional treatment train. Bacterial degradation of organics was only a factor in the pilot plant system, since the Torresdale plant inhibited bacterial growth through prechlorination.   In addition to THMFP and DOC reductions, postponing the chlorination process should be considered for another reason: Prechlorination at Torresdale appeared to produce many more nonpolar compounds (both chlorinated and nonchlorinated), as observed from the broad spectrum analysis of extracts from macroreticular resin accumulators. These compounds could arise from oxidation and/or substitution of low-molecular-weight organics and high-molecular-weight humics by chlorine.

The operational integrity of the pilot plant was maintained continuously for more than two years without any predisinfection. With a covered raw water basin, clarifier, and rapid sand filters, no large algae blooms or filter-clogging bacteria were noted, but bacterial levels were high throughout the system, and higher forms of animal life (e.g., freshwater shrimp) were occasionally observed in the rapid sand filters during the summer months. Bench-scale testing indicated that satisfactory disinfection of the pilot plant water could be achieved with postchlorination.

Ozonation Systems

The ozonation process affected the level of organic compounds (both natural and synthetic) in the Delaware River water through two mechanisms: oxidation and volatilization. To monitor this removal, TOC, UV adsorbance at 254 nm (A254), THMFP, and specific-compound chromatography were used. Stripping was the dominant mechanism for removing the specific volatile organic compounds below a C-8 hydrocarbon (as determined by the purge and trap method for both chlorinated and nonchlorinated water). Above the C-8 hydrocarbons, stripping still predominated for the chlorinated influent, but oxidation was the predominant mechanism for removing organics in the nonchlorinated influent (as determined by the XAD resin isolation method).

Many of the more hazardous organic substances found in the water supply are low-molecular-weight organic compounds that remain stable upon ozonation. Typically, these compounds have Henry's law coefficients greater than 1×10^{-3} atm-m^3/mol and may thus be easily stripped from the water. A good example of this class of volatile compound is chloroform.

Although volatilization plays a major role in removing the lower-molecular-weight organic compounds, it was of little significance to overall DOC removal. Only a small percentage ($<$ 1%) of the DOC reductions may be accounted for by the volatilization of chloroform and other easily stripped organics, as determined by performing mass balances across the contactors. Most DOC removal observed during ozonation resulted from the complete oxidation of organics to carbon dioxide.

GC profiles of volatile and semivolatile organics provided evidence of the production of many organic compounds of unknown identity and unknown health concern during ozonation. Acetone was the only organic compound confirmed by GC/MS to be produced at detectable levels during the ozonation process. Since acetone is both highly volatile and biodegradable, it was not found in high concentrations in the O_3/GAC or O_3/sand system effluents.

Prechlorination appeared to produce compounds that were more resistant to further oxidation by ozone (as measured by DOC, A254, and THMFP). This fact was most notable for THMFP. Although THMFP levels were consistently reduced when nonchlorinated water was ozonated, the ozonation of prechlorinated water caused both increases and decreases in THMFP. Averaged through the study period, ozonation increased the THMFP levels of prechlorinated water. Thus, to oxidize organics by means of ozone, the applied water should not be predisinfected with chlorine.

Adsorption Systems

GAC systems remove both milligram-per-liter concentrations of DOC and THMFP and nano- to-microgram-per-liter concentrations of trace organics. Within this framework, each of the four adsorption systems operated during the pilot-scale testing encountered varying levels of applied organic loads. In general, however, the following trend developed:

GAC(TP) > O_3/GAC(TP) > GAC(PP) > O_3/GAC(PP)

The GAC(TP) system experienced the highest organic loads, and the O_3/GAC(PP) system experienced the lowest. This trend was observed for all of the organic parameters measured, including DOC and the volatile and semivolatile organics. The pilot plant adsorption systems experienced lower applied organic loads compared with the Torresdale Plant systems for the reasons outlined above for the conventional treatment processes. Organic levels decreased through the ozonation systems because of the production of carbon dioxide, the subsequent biodegradation of the partially oxidized products, and the volatilization of trace organics. In general, preozonation affected the concentration and types of organics applied to the GAC beds less than did the changes made in conventional treatment processes.

Sand was used in this study as an inert medium control in which biological TOC removal was observed separately from the combined adsorption/biological TOC removal observed on GAC. The basis for using sand beds in this manner was developed in bench- and pilot-scale testing, which determined that the rate of biological TOC removal on sand was equivalent to that on GAC. Thus, by comparing GAC, O_3/GAC, and O_3/sand systems during steady-state DOC removal conditions, it was possible to estimate the relative contributions of adsorption and rapid biodegradation. It was therefore determined that while preozonation increased overall DOC removal, no synergistic removal of DOC by the combined effects of adsorption and biodegradation was obtained in the preozonated GAC systems.

The O_3/GAC and the GAC systems all proved capable of removing odoriferous compounds from the process water. This reduction was nearly complete and appeared to be unaffected by the length of time the carbon bed remained in service, the type of water applied, or seasonal or temporal changes. The use of ozone, either alone or in conjunction with GAC treatment, proved to be of little additional benefit in reducing the odor threshold.

Preozonation did not significantly alter the bed life of the adsorption systems when bed life was based on limiting the effluent levels of trihalomethanes, 1,2-dichloropropane (DCP), and 10 other volatile organic compounds with average influent concentrations greater than 0.1 μg/L. The effect of preozonation on the time to both initial and 100% breakthrough for these compounds varied among compounds and systems, but maintaining GAC columns on-line beyond 100% breakthrough always resulted in a chromatographic effect, in which the instantaneous effluent concentrations exceeded the influent concentrations. The chromatographic effect was frequently observed for chloroform through both the GAC and O_3/GAC systems. But the GAC systems retained the chloroform adsorbed before bed exhaustion, whereas the net adsorption of chloroform onto the O_3/GAC systems decreased nearly to zero following breakthrough. This result may have been due to the production of unidentified low-molecular-weight organics through the ozonation process; these organics then compete for the available adsorptive sites.

Cost Considerations

The results and operating parameters from the pilot-scale investigations were used as a basis for an economic feasibility study. Though both chlorinated and nonchlorinated systems were investigated, no consideration was given to any of the costs associated with modifications to the conventional treatment processes of the Torresdale plant to allow for a nonchlorinated treatment process. Table 8 lists the assumptions used in developing the costs.

Table 8.   System Design Considerations

Flow

Peak 1817 MLD (480 mgd)
Average 1067 MLD (282 mgd)

GAC Contactors	Postfilter	Filter Adsorber
EBCT, min	15	13.8
Surface loading rate, L/min/m² (gpm/ft²)	199 (4.9)	81 (2.0)
Contactor construction	common wall concrete	common wall concrete (existing)
Number of contactor pairs	25	47
GAC bed depth, m (ft)	3 (10)	1.1 (3.75)

Ozone System

Ozone dose (avg.)	2 mg/L
Contact time, min	20
Contactor construction	reinforced concrete
	(5.5 m [18 ft] deep, 2:1 length to width)

Carbon

Cost	$1.54/kg ($0.70/lb)
Loss/regeneration	7% (by weight)

Financing

Bond interest rate	12%
Amortization period	25 years
Bond issue cost	40% of capital
Inflation for O&M	10% per year

To calculate the total costs of each of the treatment schemes, three unit processes were examined separately to determine the capital and O&M costs associated with each: GAC postfiltration contactors, GAC filter adsorbers, and ozone generators and contactors. In addition, the costs associated with carbon reactivation and/or replacement were examined. The economics of each process were affected by the capital, O&M, and financial costs. All costs investigated were based on the Torresdale plant design flow of 1067 MLD (282 mgd) and an empty bed contact time of 13.8 min for sand filter replacement (filter adsorber mode) and 15 and 30 min for the post contactor mode. Figure 3 presents the total first-year costs for the advanced water treatment systems at various carbon reactivation frequencies. More specifically, total first-year cost based on the bed life necessary to maintain various DOC, THMFP, chloroform, and DCP effluent criteria were also developed.

DISTRIBUTION SYSTEM PROBLEMS

The inclusion of granular activated carbon in the treatment train has repeatedly been shown to release a variety of heterotrophic bacteria and, on occasion, total coliforms.[4] This microbial activity can be adequately controlled through disinfection of finished water prior to transmission into the

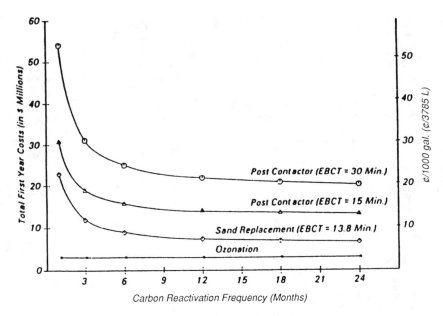

Figure 3. Advanced water treatment total cost for the first year.

distribution system. However, coliforms escaping into the distribution system cannot be ruled out because of several factors that afford protection to microbes from contact with disinfection. These factors include releases of carbon fines, passage of aggregates or clumps of bacteria from colonization sites in GAC filtration, and the protective nature of turbidity.

Preliminary findings from ongoing research on bacterial survival associated with carbon fines in drinking water that indicate carbon fines can be detected in finished water. Furthermore, carbon fines provide some measure of protection to attached *E. coli* and enterotoxigenic (heat-stable) *E. coli*. Both of these bacteria types have been isolated from carbon particles in GAC-treated drinking water. Although the association of a pathogen with carbon particles in the distribution system was found on only one occasion, it does illustrate a pathway and a reason to minimize the release of carbon fines into finished water. Since carbon fines contribute very little to nephelometric turbidity measurement (light scattering by carbon is minimal), other approaches should be considered to monitor for carbon fine releases in finished water. One approach would be to filter a standard volume of 0.5 or 1 L (0.1 or 0.3 gal) through a membrane filter and examine for particle occurrences. Another method suggested is to filter 1514 L (400 gal) of finished water through a multilayered gauze pad to search for carbon fine releases.

The water treatment train at Manchester offered a unique opportunity to study the impact of GAC filtration applied after sand filtration. This treatment configuration, in effect, placed the GAC process within one last barrier (disinfection) of water supply transmission to the distribution system.

For this reason, a variety of microbiological indicators (heterotrophic bacteria, total coliforms, acid fast bacteria, and endotoxin) were included in an 18-month study of several sampling sites on the distribution system. In addition, a variety of chemical and physical testing was done on each sample to characterize the water. Water examinations were made of the combined effluent from several GAC adsorbers, finished water, and three sampling stations on the distribution system.

Acid fast bacteria, proposed as a more restrictive indicator of disinfection effectiveness, were found to decline through the treatment train and in distribution were only recovered in densities that averaged 1 or 2 organisms/L. This result suggests that these organisms never successfully colonized either the treatment processes or the distribution system and offered no particular advantage over coliforms in measuring disinfection effectiveness.

As noted in other field studies, total organic carbon did not appear to be important in governing microbial densities found in treatment processes or in distribution water. What appears lacking is specificity to biodegradable organics, rather than total available organics that include numerous long-chain compounds not readily metabolized by microorganisms that colonize carbon filters. Likewise, the total endotoxin measurements were found to have no correlation with the microbial indicators selected. Total endotoxin relates to total biological activity, of which the indicators selected are only a small percentage of all biological forms present in the effluents or on the packed material in the GAC filter bed.

During warm water periods total coliforms were found on several occasions in finished water and distribution samples. It is interesting to note that coliforms found in the post–combined carbon filter effluents were generally inactivated by the final disinfection process, but there were two instances documented where coliforms passed into the finished water. The inconsistency of detecting coliforms in the finished water at all periods when those organisms were found in the distribution system is more a weakness of the sampling frequency of the system than a colonization event observed in very low-density coliform occurrences. The data also illustrate the value of using a 1-L sample size to examine finished water.

Another important finding was that full-scale GAC treatment resulted in a statistically significant increase in heterotrophic bacterial densities as compared to a similar plant (Concord, New Hampshire) that does not use GAC. Furthermore, water temperature, pH, and turbidity had a positive influence on heterotrophic bacterial densities.[2] These physical-chemical conditions of water are key factors that also impact disinfection effectiveness.

McFeters et al. demonstrated the problems associated with bacteria attached to granular activated carbon.[4] One of the features of this research was the establishment of a procedure to remove bacteria adsorbed to the surface of GAC to prevent reattachment, compromising bacterial viability.

Particle-associated bacteria are reported to be more resistant to disinfection. Thus, because colonized GAC filter bed particles could be released from the filter, the chlorine resistance and public health significance of these attached bacteria merit investigation.

Experiments were conducted with GAC removed from an operating drinking water filter and maintained in a column in the laboratory. Planktonic cells cultured from this column were also used. GAC-attached and planktonic cells of an *E. coli* river isolate and the pathogens *Salmonella typhimurium, Yersinia enterocolitica,* and *Shigella sonnei* were also tested. For these experiments, the attachment was accomplished by exposing virgin GAC to suspensions of each bacterial species for 20 min followed by gentle rinsing. In this time, the bacteria had little opportunity to produce extracellular material. Cells were also grown in the presence of GAC to evaluate the chlorine resistance of the GAC-attached biofilms.

Scanning electron micrographs of these particles and of those from the drinking water filter revealed colonies of bacteria and the presence of extracellular polymer. The exposure of planktonic heterotrophic plate count (HPC) cells to chlorine (2.0 mg/L) resulted in a rapid decrease in viability (5 min), whereas GAC-associated cells experienced little decline in numbers after exposure for 1 hr (Figure 4). No decrease in viability was observed within 1 hr for the GAC-grown coliforms. Some injury did occur with GAC-attached cells, suggesting that the extracellular polymer produced by the grown cells or the integrity of the attached colony provided some amount of protection from chlorine.

These data suggest a means by which bacteria, including pathogens, can breach disinfection barriers and enter distribution systems.

SUMMARY AND CONCLUSIONS

Granular activated carbon filters and GAC adsorbers used in water supply treatment have been investigated in a variety of treatment train configurations used in pilot-plant systems and in full-scale treatment operations. Microbial activity does occur in GAC processes with regrowth related to biodegradable organics and warm water periods. Bacterial populations in GAC filter beds can be minimized through application of disinfection to the influent waters. All carbon materials used for organic removal can become colonization sites for some bacteria in water passing through the filter bed. This material may also become contaminated from bulk storage in unprotected areas, placement in converted sand filter beds not adequately cleaned, or transfer of GAC in contaminated slurry to filter beds. Reactiva-

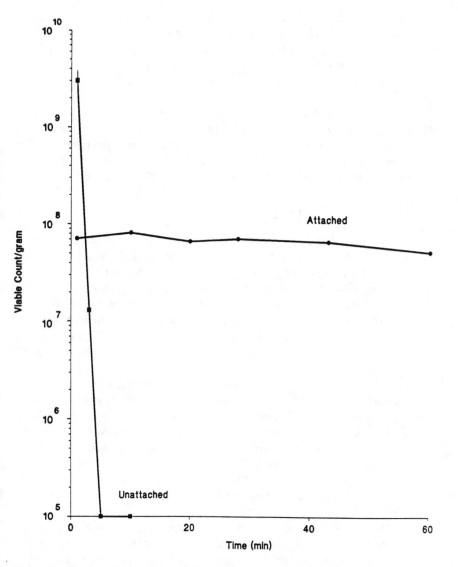

Figure 4. Survival of naturally occurring heterotrophic plate count bacteria exposed to chlorine at 2.0 mg/L for 1 hr. (Free chlorine residual after 1 hr was 1.7 mg/L.)

tion of GAC may cause some particle size reduction and change in surface characteristics that appear to have little effect on the ability of heterotrophic bacteria to find sites for attachment and colonization. Carbon fines in GAC material may be a transport vehicle for bacteria during the stabilization period of a new bed with virgin GAC or filter replaced with reactivated GAC.

The higher velocity of process water passing through pressure contactors or adsorbers does minimize the continued presence of high densities of heterotrophic bacteria in the GAC effluent after a stabilization of these systems is realized. Total coliform regrowth also appears to be minimized in these closed devices. Filter adsorbers are impacted by floc or silt that coats the carbon particles. This condition contributes to shorter usefulness of the adsorption process and to the potential for colonization by coliforms with their subsequent release into the GAC process effluent. Connecting contactors in series in one pilot study indicates coliform passage from one contactor to another can occur with releases in the final effluent. However, colonization was not permanently established in any of the four contactors in series.

Careful consideration must be given to placement of the GAC process in the water treatment train as it relates to optimizing organic removal, GAC bed life, and controlling microbial resurgence by effective disinfection. Stabilization of the GAC bed after backwashing should be achieved quickly to avoid significant passage of floc, silt, carbon fines, and bacterial aggregates through the final stages of treatment. Using chlorinated backwash water, more frequent backwashing, and discharging the startup GAC filter effluent to waste for the first 10 min is desirable. Application of disinfection to GAC filter influents and prior sand filtration will optimize the production of effluents that will not contain coliform bacteria in 100-mL test portions.

The effects that GAC treatment has on distribution water quality are largely unknown. In theory, the long-term benefit to improve microbial quality of water in distribution systems would be a reduction in available nutrients (organics) to support bacterial regrowth in the pipe sediments and tubercles, provided GAC is replaced or reactivated frequently. Since GAC treatment does result in significant bacterial activity farther into the treatment train, more attention should be given to monitoring finished water for the appearance of carbon fines and coliform occurrences in liter-volume test samples.

A pathway involving passage of carbon particles, coliform attachment, and protection from disinfection inactivation has not been clearly demonstrated in published case history reports on coliform problems in distribution networks. However, it should be noted that some of the water systems involved used carbon-capped filters for GAC filtration. Investigation of bacterial quality in a water system using GAC treatment did reveal, on one occasion, evidence of a protected pathway for coliform passage from carbon effluent through final disinfection then out into the distribution system. These coliform occurrences were based on three replicate examinations of 1-L samples and illustrate how very low densities of coliforms go undetected when only 100-mL sample volumes are analyzed.

An extensive review of bacteriological data from a variety of pilot-plant and full-scale operations using GAC treatment was made. Major items included GAC characteristics, sanitary quality of virgin carbon, GAC reactivation effects, gravity filtration vs pressure contactors, significance of GAC placement in the water treatment train, and the potential impact on microbial quality of distribution water. None of these issues is beyond control using reasonable treatment precautions by water plant operators. However, there is a need to revise the monitoring of finished water for detection of carbon particle releases and coliform densities at 1-L levels to provide more useful information to fine-tune treatment effectiveness.

REFERENCES

1. Schulof, P. "An Evolutionary Approach to Activated Carbon Treatment in France." *J. Environ. Pathol. Toxicol. Oncol.* 7(718):55–75 (1987).
2. Lykins, Jr., B. W., E. Geldreich, J. Q. Adams, J. Ireland, and R. M. Clark. "Field Scale Granular Activated Carbon Studies for Removal of Organic Contaminants Other than Trihalomethanes from Drinking Water," U.S. EPA Report 600/2-84-165 (1984).
3. Neukrug, H. M., M. G. Smith, J. T. Coyle, J. P. Santo, J. McElhaney, I. H. Suffit, S. W. Maloney, P. C. Chrostowski, W. Pipes, J. Gibs, and K. Bancroft. *Removing Organics from Philadelphia Drinking Water by Combined Ozonation and Adsorption* (Cincinnati, OH: U.S. EPA, Office of Research and Development, WERL).
4. McFeters, G. A., A. K. Cumper, M. W. LeChevallier, S. C. Broadway, and D. G. Davies. "Bacteria Attached to Granular Activated Carbon in Drinking Water," U.S. EPA Report 600/M-87/003 (June 1987).

CHAPTER 9

Adsorption Modeling

INTRODUCTION

Many investigators have explored the development of mathematical models for describing the kinetics of adsorbate removal in fixed granular activated carbon beds. One of the more widely used mathematical models for GAC adsorption is the homogeneous surface diffusion model (HSDM) developed by Weber, Crittenden, and co-workers.[1-4] The following assumptions are made in this fixed bed model:

1. The hydraulic loading and influent concentration are constant.
2. There is no radial dispersion or channeling.
3. Surface diffusion flux is much greater than pore diffusion flux as an intraparticle mass transfer mechanism.
4. The liquid phase diffusion flux can be described by the linear driving force approximation, using estimates for the film transfer coefficient k_f.
5. The adsorbent is fixed in the adsorber.
6. Adsorption equilibrium can be described by the Freundlich isotherm.
7. Plug flow exists within the bed.

Crittenden and Hand et al. developed solutions to the HSDM for fixed bed adsorber systems to aid the design engineer in evaluating adsorber performance and determining least-cost adsorber operations.[1,5] Hand et al. presented both analytical and numerical solutions for the HSDM and Crittenden and co-workers have presented a solution to the HSDM based on the use of orthogonal collocation techniques.[1,5] Adsorption modeling is discussed in more detail in the following sections.

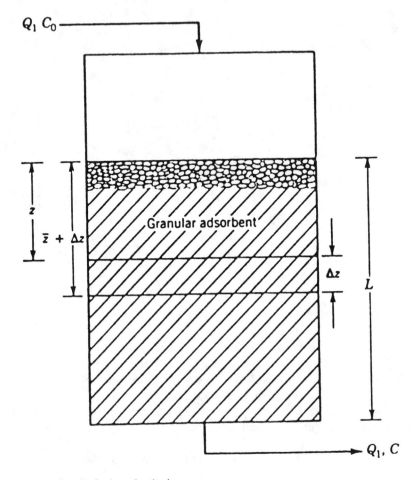

Figure 1. Detail of adsorption beds.

ADSORPTION MODELING

A downflow contactor system is shown schematically in Figure 1. The adsorbent is packed granular media fixed in a stationary position. Process water is applied directly to one end and forced through the bed, typically by gravity, pressure, or pumping energy. Depending on the characteristics of the adsorbent as well as the physical and hydraulic nature of the system, all the adsorbate may be removed before the process water appears in the effluent. The region of the bed where adsorption and removal of the adsorbate takes place is referred to as the mass transfer zone. As the initially contacted adsorbent is exhausted, the mass transfer zone moves further into the bed in a wavelike manner; accordingly, the mass transfer zone is sometimes referred to as the adsorption wave. When the entire bed becomes

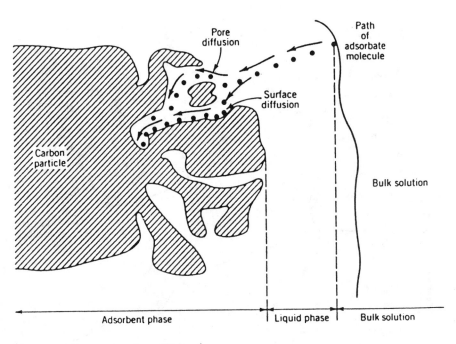

Figure 2. Pore and surface diffusion.[1]

exhausted and the mass transfer zone approaches the end of the bed, increasingly higher concentrations of the adsorbate are observed, until the effluent concentration equals influent concentration and no removal occurs. This phenomenon is termed breakthrough. In practice, the bed would be replaced with fresh adsorbent or switched with an alternate unit once a predetermined treatment goal of the adsorbate is attained.

As shown in Figure 2, the adsorbate can diffuse by two mechanisms within the adsorbent, pore diffusion and surface diffusion. For pore diffusion, the adsorbate is transported within the pore fluid. For surface diffusion, the adsorbate continues to move along the surface of the adsorbent to available adsorption sites as long as it has enough energy to leave its present site. Investigations have demonstrated that surface diffusion is the dominant mechanism, so the contribution of pore diffusion is neglected.[2-6] The assumption of adsorbent homogeneity is a simplification of the complex inner pore and surface structure.

Two of the three basic differential equations of the model are mass balances for the adsorbate in the axial direction, z, shown in Figure 1, and in the radial direction, r, shown in Figure 3. The liquid phase mass balance describes the spatial and temporal variation of the adsorbate concentration in solution. The solid phase mass balance describes the removal rate of

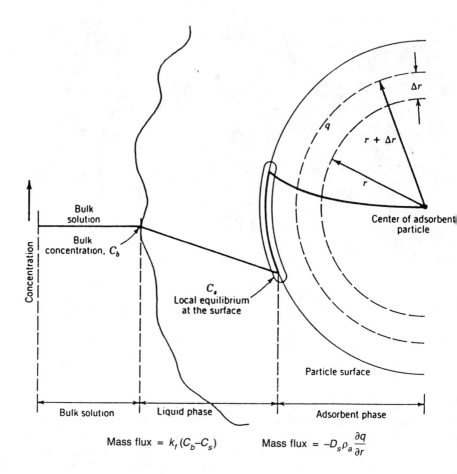

Mass flux $= k_f(C_b - C_s)$

Mass flux $= -D_s \rho_a \dfrac{\partial q}{\partial r}$

Figure 3. Mass balance for adsorbate.[1]

adsorbate from the liquid phase into the adsorbent at a given axial location in the bed.

A mass balance for the adsorbate in the liquid phase system defined in Figure 3 can be written as:

$$r_a = r_{fin} - r_{fout} - r_r$$

where r_a = rate of accumulation of adsorbate within the longitudinal element

r_{fin} = rate of flow of adsorbate into longitudinal shell by advection

r_{fout} = rate of flow of adsorbate out of longitudinal shell by advection

r_r = rate of removal of adsorbate by adsorbent

Mathematical Models

A large number of mathematical models have been developed to describe the adsorption of organics in GAC columns. Models have been developed to the point of including variable flow rates, organic concentrations, and multicomponent mixtures of sorbates.[5-10] Sorption models nearly always include two basic steps: mass transfer from the bulk liquid to the particle surface and intraparticle transport. The adsorption step is generally considered to be very rapid and local equilibrium conditions are assumed to exist. Granule geometry is usually normalized to a spherical shape, although shape factors are often included. Transport of the sorbate from the bulk liquid to the granule surface is generally assumed to be controlled by a stagnant film resistance and a liner mass transfer model is used to describe the process as follows:

$$\epsilon \Delta V \frac{\partial C}{\partial t} = A_g k_f (C_i - C_s) \tag{1}$$

where ϵ = porosity of bed
  ΔV = increment of bulk bed volume, m^3
  C_i = concentration of species in the bulk liquid, g/m^3
  C_s = concentration of the species at granule surface, g/m^3
  A_g = surface area of granule, m^2
  k_f = mass transfer coefficient, m/sec

To couple the solid and liquid phase mass balance equations, the surface concentration of adsorbate in the liquid phase $C_s(t)$ must be expressed in terms of the surface concentration of adsorbate in the solid phase q (R,t). This equation is developed from the assumption of local adsorption equilibria adjacent to the exterior adsorbent surface, as described in the nonlinear Freundlich isotherm:

$$q = K C_s(t)^{1/n} \quad (r = R, t) \tag{2}$$

Intraparticle transport of organics dissolved in water has been shown to involve two mechanisms, pore diffusion and surface diffusion.[11-13] Models can be constructed that combine the two mechanisms using an effective diffusivity, D_s.[14,15] Such models require calibration, and like isotherms cannot be expected to be valid outside the calibration region. Pore and surface diffusion coefficients usually have values in the range of 10^{-8} to 10^{-10} and 10^{-13} to 10^{-15} m^2/sec respectively.[5,14,16] The effective diffusivity can be calculated as a sum of the pore and surface diffusivities using conversion factors to relate surface and pore conditions:[14]

$$D_e = D_p + K\rho_a D_s \tag{3}$$

where D_e = effective diffusivity, m^2/sec
      D_p = pore diffusivity, m^2/sec
      K = liquid-solid distribution coefficient, m^3/g
      ρ_a = density of GAC granule including pore volume, g/m^3
      D_s = surface diffusivity, m^2/sec

The coefficients K and ρ_a effectively relate the quantity of sorbed material to the concentration in the pore and allow the development of a combined effect diffusion equation. A nonlinear adsorption relationship (n $\neq$ 1) results in a more complex relationship for Equation 3 but the concept is the same. Assuming spherical particles and variation only in the radial direction in addition to the assumptions stated above results in the Fick's Law model given in Equation 4:

$$\frac{\partial C_r}{\partial t} = D_e \frac{1}{r^2 \partial r} r^2 \frac{\partial C_r}{\partial r} \tag{4}$$

where C_r is the solute concentration (m^2/s) at a distance r from the center of the particle and t is elapsed time.

Expressions similar to Equation 4 can be independently written for the pore and solid diffusion processes. Integration of Equation 4 or the analogous expressions for pore diffusion over the distance from the surface to the center of the granules requires consideration of the actual length of travel due to the twisting of the pores. This effect is taken into account by including a parameter called the tortuosity, τ, the ratio of the true length of path to the radial distance. Also included in the tortuosity is the effect of blind pores. Values of the tortuosity are usually between 2 and 5.[14,15]

Crittenden and co-workers have concluded that in most cases pore diffusion is a small factor in intraparticle transport and suggest a form of Equation 4 based solely on surface diffusion.[5-10]

$$\frac{\partial q_r}{\partial t} = D_s \frac{1}{r^2 \partial r} r^2 \frac{\partial q_r}{\partial r} \tag{5}$$

where q_r is the phase concentration (g/kg) of an adsorbate at a distance r from the center of the granule, and t is time.

Elimination of pore diffusion in Equation 5 is particularly helpful because the Freundlich exponent, $1/n$, is commonly reported in the range of 0.2 to 0.8 and the resulting combined pore and surface diffusion model is difficult to solve.[17]

Crittenden and co-workers have developed models based on Equations 1 and 5, and a coupled axial transport model for the bulk fluid flow region that have been very useful in analyzing adsorption processes and mechanisms. An important feature of the models is that the presence of contaminant mixtures is accounted for using ideal adsorbed solution theory (IAST), which is based on an assumption of dilute conditions and noninteraction between sorbate species.[18] The requirements of IAST appear to be met for most situations where VOCs are being removed from drinking water. However, the presence of humic substances presents problems in characterization of the multicomponent effects. The models require calibration; that is, coefficients cannot be developed completely independently of experimental studies with the water and GAC to be used in the prototype systems. In applying these models, evaluation of the coefficients is very important. The film mass transfer coefficient, k_f, is estimated using correlations with the hydraulic loading rate, Schmidt and Reynolds number. The surface diffusion coefficient is obtained experimentally from systems in which film mass transfer limitations have been minimized or through application of Equation 6:

$$D_s = PSDFR \left(\frac{D\epsilon_p C_i}{\tau K C_i^{1/n} \rho_a} \right) \tag{6}$$

where PSDFR = ratio of the surface to pore diffusion flux contribution
D_s = liquid diffusion coefficient of adsorbate, m²/sec
ϵ_p = pore void fraction

Other terms have been defined previously.

The bracketed term in Equation 6 is termed the pore diffusion flux contribution (PDFC). All of the terms in the PDFC, other than the tortuosity, can be determined independently. The tortuosity can be estimated using mean pore diameters and the void fraction. Reported values of the PSDFR are between 3 and 10.[5]

A number of dimensionless numbers and concepts, useful in applying these equations, are presented in the following section.

Dimensionless Numbers

Mass throughput or dimensionless time is defined as:

$$T = \frac{\text{rate of mass of adsorbate fed}}{\text{total mass of adsorbate at equilibrium}} \tag{7}$$

$$= \frac{QC_i t}{Mq_e + \epsilon V C_i} \tag{8}$$

where Q = influent flow rate

C_i = fluid phase concentration of adsorbate in influent

t = elapsed time

M = total mass of adsorbent in the bed

q_e = adsorbent phase concentration at equilibrium with C_i in fluid phase

ϵ = ratio of void volume to total bed volume

V = total bed volume

The dimensionless solute distribution parameter D_g is defined as:

$$D_g = \frac{\text{mass of adsorbate in solid phase at equilibrium}}{\text{mass of adsorbate in liquid phase at equilibrium}} \qquad (9)$$

$$= \frac{\rho_a q_e (1-\epsilon)}{\epsilon C_i} \qquad (10)$$

where ρ_a = adsorbent density including pore volume.

The dimensionless Biot number, B_i, is defined as:

$$B_i = \frac{\text{rate of liquid phase mass transfer}}{\text{rate of solid phase mass transfer}} \qquad (11)$$

$$= \frac{(1-\epsilon)k_f R}{\epsilon D_s D_g \Phi} \qquad (12)$$

where k_f = film transfer coefficient

D_s = surface diffusion coefficient

Φ = sphericity (dimensionless ratio of the surface area of the equivalent volume sphere to the actual surface area of the particle)

R = particle radius

The modified Stanton number, St, is a dimensionless measure of the bed length as compared to the length of the mass transfer zone in the case where liquid phase mass transfer resistance controls the adsorption rate:

$$St = \frac{k_f EBCT(1-\epsilon)}{R\epsilon\Phi} \qquad (13)$$

where EBCT = hydraulic residence time in the bed or empty bed contact time. The surface diffusion modulus E_d is a dimensionless measure of bed length compared to the length of the mass transfer zone in the case where intraparticle diffusion controls adsorption rate:

$$E_d = \frac{D_s D_g \tau}{R^2} = \frac{St}{B_i} \tag{14}$$

In addition to the theoretical approach as described by Crittenden and co-workers, there are other more empirical approaches that have been effective. An approach that has been useful and has been applied to field-scale systems is described in the following section.

FIELD-SCALE SYSTEMS

Information from the Cincinnati Water Works and Jefferson Parish will be analyzed in this section using an empirical model that was developed for characterizing total organic carbon breakthrough curves from GAC.[19-21] The model will be applied to pilot-column data from both projects and compared to the results from field-scale equivalents. Data from the Jefferson Parish project will then be applied to scale-up projections. A model based on best current cost estimates of "in-place" technology combined with the performance model will be used to estimate a minimum cost design for a full-scale system. The Cincinnati and Jefferson Parish systems have been discussed previously.

Model Development

The model developed in this section is based on the use of a mass transfer concept in combination with the Freundlich isotherm. GAC adsorption systems used in drinking water treatment typically use stationary beds with the liquid flowing downward through the adsorbent. Under these conditions, adsorbed material accumulates at the top of the bed until the amount adsorbed at that point reaches a maximum.[21] The maximum amount of a contaminant that can be adsorbed on activated carbon occurs when the adsorbed material is in equilibrium with the concentration of the contaminant solution surrounding the adsorbent. For any given concentration of material in the liquid phase, the loading on or "capacity" of activated carbon at equilibrium can be determined from the appropriate "adsorption isotherm." When the adsorbed material is in equilibrium with the influent concentration, the bed is "loaded" to capacity, and that portion of the bed is "exhausted." In an ideal "plug flow" situation, the exhausted zone moves downward with time in service until the entire adsorbent bed is exhausted.

Writing a liquid phase mass balance within a differential element in a fixed bed reactor results in Equation 1, where the liquid phase contaminant mass flow rate in, less the liquid phase contaminant flow rate out, divided by the differential element volume, equals the mass rate of contaminant transferred to the solid phase. Equation 15 is as follows:

$$J = \frac{[G_sEC - G_sE(C - \Delta C)]}{E(\Delta Z)} \tag{15}$$

where J = mass transfer rate per unit reactor volume, mg solute/hr/m^3
   G_s = flow rate of solvent per unit of cross-section area, m^3/hr/m^2
   E = column cross-section area, m^2
   C = influent contaminant concentration, mg/m^3
   ΔC = incremental change in C, mg/m^3
   ΔZ = differential reactor height, m

Cancelling E and G_sEC and letting $\Delta C / \Delta Z \to dC/dZ$ yields:

$$J = (G_s) \, dC/dZ \tag{16}$$

Using the mass transfer coefficient concept within a differential height of adsorber dZ yields:[22,23]

$$K_T (C - C_e) = G_s \, dC/dZ \tag{17}$$

where K_T = mass transfer coefficient, hr^{-1}
   C_e = equilibrium value for contaminant at the solid surface, mg solute/m^3

Based on an assumption that all the solvent is removed at the bottom of the column, an ideal mass balance over the entire column is:[23]

$$G_sC = L_aX \tag{18}$$

where G_s = flow of solvent, m^3/hr/m^2
   L_a = mass velocity of adsorbent to keep mass transfer zone stationary, mg/hr/m^2 (as described in Traybal[23])
   X = concentration of adsorbed solute per unit weight of adsorbent, mg solute/mg adsorbent

If all the solvent is removed at the bottom of the column, then:

$$\frac{L_a}{G_s} = \frac{C}{X} \tag{19}$$

The rate of solute transfer over the differential height of adsorber (dZ) can be written as in Equation 15, where C_e is the equilibrium composition in the liquid corresponding to the adsorbate composition X and $C - C_e$ is the driving force to equilibrium.

From Equation 15:

$$\frac{dC}{C - C_e} = \frac{K_T dA}{G_s} \tag{20}$$

Using the Freundlich isotherm between the adsorbent and solvent yields:

$$X = K_e(C_e)^{1/n} \tag{21}$$

where K_e = equilibrium constant
$1/n$ = slope of isotherm

From Equation 21:

$$C_e = (1/K_e)^n X^n \tag{22}$$

Substituting Equations 19 and 22 into Equation 20 yields:

$$\frac{dC}{C-[(1/K_e)^n(G_s/L_s)^n]C^n} = (K_T/G_s)dZ \tag{23}$$

Let the velocity of the adsorption zone be:[24]

$$V = dZ/dt \tag{24}$$

and let

$$K = (1/K_e)^n(G_s/L_a)^n \tag{25}$$

and let

$$R = (K_T/G_s)V \tag{26}$$

Substituting Equations 24–26 into Equation 21 yields:

$$\frac{dC}{C-KC^n} = Rdt \tag{27}$$

The general integral between C_b (concentration at breakthrough) and C and t_b (time at breakthrough) and t for Equation 27 yields:

$$\int_{C_b}^{C} \frac{dC}{C-KC^n} = R \int_{t_b}^{t} \frac{t}{} dt \tag{28}$$

Integrating Equation 28 yields:

$$\frac{1}{(n-1)}\ln\left[\frac{C^{n-1}(1-KC_b^{n-1})}{C_b^{n-1}(1-KC^{n-1})}\right] = R(t-t_b) \tag{29}$$

or

$$\frac{C_b^{n-1}}{KC_b^{n-1} + (1-K\,C_b^{n-1})e^{-R(n-1)(t-t_b)}} = C^{n-1} \tag{30}$$

If $R(n-1) = r$, then

$$\frac{1/K}{1 + [\frac{1}{KC_b^{n-1}} -1]\, e^{rt_t^{-rt}}} = C^{n-1} \tag{31}$$

As $t \to \infty$, $1/K \to C_i^{n-1}$ where C_i is the constant influent value on the carbon bed.

Equation 31 can be rewritten as:

$$\left[\frac{C_1^{n-1}}{1 + [(\frac{C_i^{n-1}}{C_b^{n-1}} -1) \, e^{rt}{}_b] \, e^{-rt}} \right]^{1/n-1} = C \tag{32}$$

or

$$\left[\frac{C_1^{n-1}}{1 + Ae^{-rt}} \right]^{1/n-1} = C \tag{33}$$

where

$$A = [\frac{C_i^{n-1}}{C_b^{n-1}} -1] \, e^{rt}b \tag{34}$$

Equation 33 is the generalized logistic function.[24,25]

Validation of Model

Equation 33 was applied to field-scale data to illustrate the comparison of pilot and full-scale systems located at the Jefferson Parish and Cincinnati projects for TOC removal. Isotherm relationships were estimated based on effluent histories taken from pilot and field-scale contactor experiments at the two sites. The carbon adsorbed in mg/g at column exhaustion was plotted versus average influent concentration in mg/L to obtain constants in Equation 21. The Freundlich isotherm relationships for Jefferson Parish and Cincinnati, respectively, are as follows:

$$q = 8.21 \, C_e^{0.316} \qquad (r^2 = 0.70) \tag{35}$$

and

$$q = 90.11 \, C_e^{-0.883} \qquad (r^2 = 0.82) \tag{36}$$

where q = equilibrium concentration of TOC adsorbed, mg/g
C_e = equilibrium concentration of TOC in solution

To illustrate the comparison between full-scale and pilot columns, Equation 33 was fit to pilot column data as shown in Figure 4. The dashed line in Figure 4 is the best fit of Equation 33 to the pilot column effluent data,

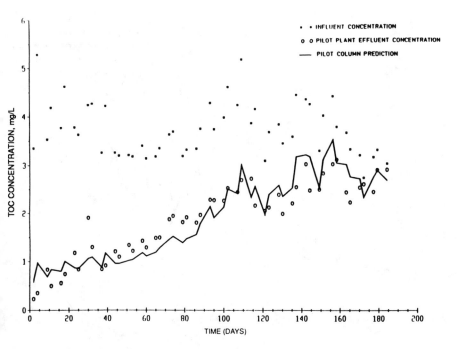

Figure 4. Comparison of predicted and actual pilot column TOC concentration, Jefferson Parish, Louisiana.

assuming the influent concentration varies with time. The resulting equation is as follows:

$$C = \left[\frac{C_i^{2.16}}{1 + 42.71 \; e^{-0.27t}} \right]^{0.463} \tag{37}$$

Figure 5 shows the plot of Equation 37, the effluent from the filter adsorber, and the influent history common to both the filter adsorber and the pilot unit adsorber.

As can be seen from Figure 5, the pilot unit equation is a good predictor of the full-scale system. An alternative approach would be to estimate the parameters in Equation 33 for both pilot and full-scale units and perform a significance test to quantitatively determine how well the pilot predictions fit the full-scale unit. For purposes of this discussion, a qualitative comparison is adequate.

Figure 6 shows the adsorber pilot column prediction, the effluent, and the influent history for the full-scale adsorber. In this case, the pilot column appears to be an even better predictor of the full-scale system than of the filter adsorber. The best-fit equation is as follows:

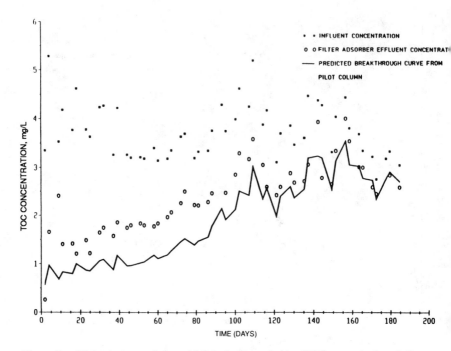

Figure 5. Pilot column prediction of full-scale filter adsorber TOC concentration, Jefferson Parish, Louisiana.

$$C = \left[\frac{C_i^{2.16}}{1 + 80.25\ e^{-0.0029t}} \right]^{0.463} \tag{38}$$

Figures 7 and 8 show the same pilot and full-scale comparison at the Cincinnati project for the filter adsorber and adsorber for removal of TOC. The respective equations are as follows:

$$C = \left[\frac{C_i^{-2.13}}{1 + 0.96\ e^{-0.0077t}} \right]^{-0.469} \tag{39}$$

and

$$C = \left[\frac{C_i^{-2.13}}{1 + 0.99\ e^{-0.0019t}} \right]^{-0.469} \tag{40}$$

It is obvious from this analysis that pilot columns can adequately predict the performance of full-scale units.

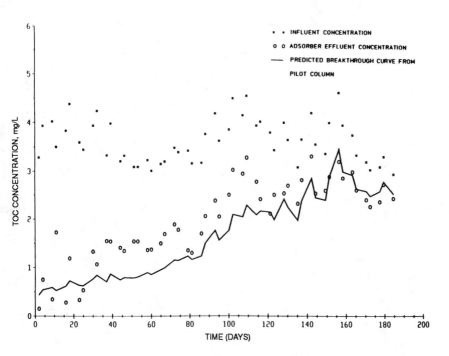

Figure 6. Pilot column prediction of full-scale adsorber TOC concentration, Jefferson Parish, Louisiana.

Simulation of Full-Scale System

To simulate a full-scale system, data from the four pilot columns at Jefferson Parish with empty bed contact times of 10, 20, 30, and 40 min, respectively, were used. Figure 9 shows the influent and effluent histories, along with the best fit lines for these four pilot columns. The best-fit equations are as follows:

$$C_{10} = [C_i^{2.16}/(1 + 16.25e^{-0.026t})]^{0.461} \tag{41}$$

$$C_{20} = [C_i^{2.16}/(1 + 51.07e^{-0.024t})]^{0.461} \tag{42}$$

$$C_{30} = [C_i^{2.16}/(1 + 61.67e^{-0.016t})]^{0.461} \tag{43}$$

$$C_{40} = [C_i^{2.16}/(1 + 118.28e^{-0.015t})]^{0.461} \tag{44}$$

As can be seen from Equations 41 through 44, the values for A and r vary with EBCT. Plotting A and r versus EBCT yields Figures 10 and 11, which show best-fit equations relating A and r to EBCT(T).

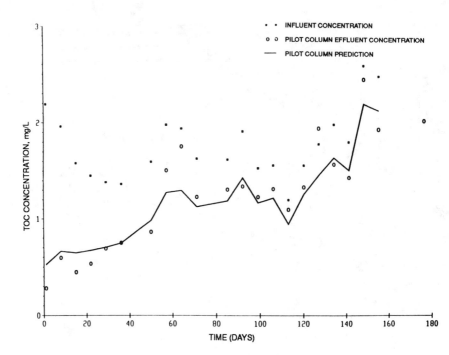

Figure 7.   Comparison of filter adsorber predicted and actual pilot column TOC concentrations, Cincinnati, Ohio.

By choosing EBCTs below 10 min and above 40 min on the lines in Figures 10 and 11, values for A and r can be selected to extrapolate Equation 31. For example, by choosing an EBCT of 5 min, the values of A = 6.67 and r = 0.035 are selected using Figures 10 and 11. These values can be used to provide an estimated breakthrough curve for 5-min EBCT. Extrapolation was used because there are no data available for shorter or longer EBCTs. A preferable approach would be to have actual column experiments for a wider range of EBCTs. The problems and limitations of using extrapolated data are discussed later. Table 1 contains the parameters for EBCTs of 1, 5, 10, 20, 30, 40, 50, 60, 70, and 80 min.

Solving each of the resulting equations for breakthrough times at levels of 0.5, 0.75, 1.0, 1.5, and 2.0 mg/L yields the results in Table 2. These values are based on the assumption of a constant influent value of C_i = 3.5 mg/L.

The values in Table 2 can be converted to a carbon use rate of kg/day (lb/day) for various effluent targets by estimating the intersection of the breakthrough target with the breakthrough curve (weight per column was 1.28 kg [7.85 lb]). This intersection yields a breakthrough period, which when divided into the equivalent weight for EBCT yields a carbon use rate.

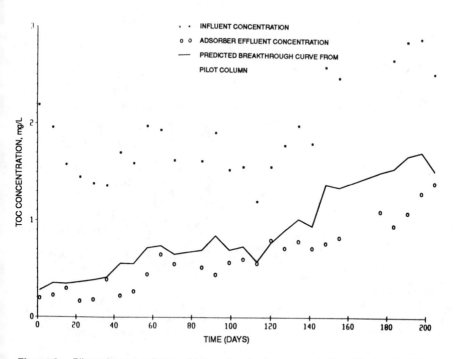

Figure 8. Pilot column prediction of full-scale adsorber concentration, Cincinnati, Ohio.

Table 3 contains the use rates in lb/day for various EBCTs and target levels associated with the pilot columns and the extrapolated values.

Projection to Full-Scale System

To illustrate the application of these data to a full-scale configuration, the following conditions will be assumed: a TOC influent concentration of 3.5 mg/L; a design capacity of 0.26 m³/sec (70 mgd); an average flow of 0.21 m³/sec (55 mgd); 20 contactors, with an average cross-section of 9.91 m (32.5 ft) × 9.91 m (32.5 ft); and a hydraulic loading of 0.9 m/hr (2.3 gpm/ft²) for a full-scale system design. No attempt was made to estimate the effect of an increased loading rate on breakthrough. Using these data and the pilot column use rate data from Table 3 and assuming direct proportionality, the GAC use rates in lb/day for the full-scale system were calculated as shown in Table 4.

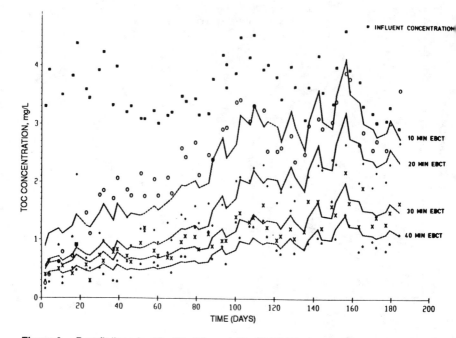

Figure 9. Best-fit lines for 10-, 20-, 30-, and 40-min EBCTs from pilot column, Jefferson Parish, Louisiana.

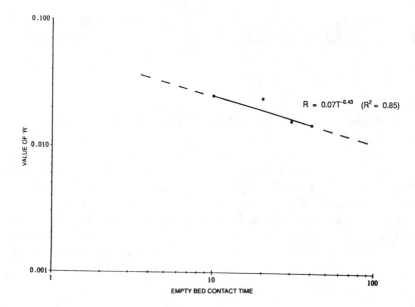

Figure 10. Natural log of r vs natural log of EBCT.

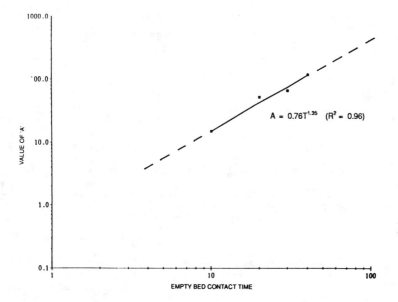

Figure 11.   Natural log of A vs natural log of EBCT.

Table 1.   Parameters for Generalized Logistic Function

EBCT (min)	A	r
1	0.76	0.070
5	6.67	0.035
10	17.01	0.026
20	43.37	0.019
30	74.98	0.016
40	110.56	0.014
50	149.42	0.013
60	191.25	0.012
70	235.34	0.011
80	281.83	0.011

Table 2.   Breakthrough Time in Days

EBCT (min)	0.5	0.75	1.0	1.5	2.0
1	—	—	—	—	—
5	—	—	—	6.7	29.6
10	—	—	7.1	44.9	75.8
20	—	24.3	58.9	110.8	153.0
30	6.8	63.2	104.2	165.7	215.9
40	35.6	99.9	146.8	217.2	274.5
50	61.4	130.8	181.3	257.0	318.8
60	87.1	162.2	216.9	299.0	366.0
70	113.9	195.8	255.5	345.0	418.1
80	130.3	212.2	271.5	361.4	434.5

Table 3.   Carbon Use Rate in lb/day for Pilot Columns[a]

EBCT (min)	TOC Target Level				
	0.5	0.75	1.0	1.5	2.0
1	—	—	—	—	—
5	—	—	—	0.59	0.13
10	—	—	1.11	0.17	0.10
20	—	0.65	0.27	0.14	0.10
30	3.46	0.37	0.23	0.14	0.11
40	0.88	0.31	0.21	0.14	0.11
50	0.64	0.30	0.22	0.15	0.12
60	0.54	0.29	0.22	0.16	0.13
70	0.48	0.28	0.22	0.16	0.13
80	0.48	0.30	0.23	0.17	0.14

[a]To convert from lb/day to kg/day, multiply by 0.4535.

Table 4.   Full-Scale GAC Use Rates (lb/day)[a]

EBCT (min)	TOC Target Level				
	0.5	0.75	1.0	1.5	2.0
5	—	—	—	143402.8	31597.228
10	—	—	269791.72	41319.452	24305.56
20	—	107986.14	65625.012	34027.784	28437.505
30	840972.38	89930.572	55902.788	34027.784	26736.116
40	213888.93	75347.236	51041.676	34027.784	26736.116
50	155555.58	52916.68	53472.232	36458.34	29166.672
60	131250.02	70486.124	53472.232	38888.896	31597.228
70	116666.69	68055.568	53472.232	38888.896	31597.228
80	116666.69	72916.68	55902.788	41319.452	34027.78

[a]To convert from lb/day to kg/day, multiply by 0.4535.

REFERENCES

1. Hand, D. W., J. C. Crittenden, and W. E. Thacker. "User-Oriented Solutions to the Homogeneous Surface Diffusion Model for Adsorption Process Design Calculations, Part 1, Batch Reactor Solutions," paper presented at the 54th Annual Conference of the Water Pollution Control Federation, Detroit, Michigan (October 4–9, 1981).
2. Brecher, L. E., D. C. Frantz, and J. A. Kostecki. "Combined Diffusion in Batch Adsorption Systems Displaying BET Isotherms: Part II." *AIChE Symp. Ser.* 63(74):25–30 (1967).
3. Furusawa, T., and J. M. Smith. "Fluid-Particle and Intra-Particle Mass Transport Rates in Slurried Beds, *Ind. Chem. Fund* 12(2):197(1973).
4. Komiyama, H., and J. M. Smith. "Surface Diffusion on Liquid-Filled Pores," *AIChE J.* 20(6):1110 (1974).
5. Crittenden, J. C., D. W. Hand, H. Arora, and B. J. Lykins. "Design Considerations for GAC Treatment of Organic Chemicals," *J. Amer. Water Works Assoc.* 79:77 (1987).

6. Crittenden, J. C. "Mathematical Modeling of Adsorber Dynamics: Single Components and Multi-Components," Ph.D. Thesis, University of Michigan, Ann Arbor, MI (1976).

7. Crittenden, J. C., and W. J. Weber, Jr. "Predictive Model for Design of Fixed Bed Adsorbers: Parameter Estimation and Model Development," *J. Environ. Eng. Div. ASCE* 104:185 (1978).

8. Crittenden, J. C., and W. J. Weber, Jr. "Model for Design of Multi-Component Adsorption Systems," *J. Environ. Eng. Div. ASCE* 104:1175 (1978b).

9. Crittenden, J. C., B. W. C. Wong, W. E. Thacker, V. L. Snoeyink, and R. L. Hinrichs. "Mathematical Modeling of Sequential Loading in Fixed Bed Adsorbers," *J. Water Pollut. Control. Fed.* 52:2780–2795 (1980).

10. Crittenden, J. C., P. Luft, D. W. Hand, J. L. Oravitz, S. W. Loper, and M. Ari. "Prediction of Multicomponent Adsorption Equilibria Using Ideal Adsorber Solution Theory," *Environ. Sci. Tech.* 19:1037 (1985).

11. Hiester, N. K., and T. Vermeulen. "Saturation Performance of Ion Exchange and Adsorption Columns." *Chem. Eng. Prog.* 48:505 (1952).

12. Vermeulen, T. "Theory for Irreversible and Constant Pattern Solid Diffusion," *Ind. Eng. Chem.* 45:1664 (1953).

13. Vermeulen, T. "Separation by Adsorption Methods," in *Advances in Chemical Engineering*, Vol. II, T. B. Drew and J. W. Hoopes, Eds. (New York: Academic Press, 1958).

14. Jackman, A. P. and K. T. Ng. "The Kinetics of Ion Exchange on Natural Sediments," *Water Resour. Res.* 22:1664 (1986).

15. Summers, R. S., and P. V. Roberts. "Simulation of DOC Removal in Activated Carbon Beds," *J. Env. Eng. Div., ASCE* 110:73 (1984).

16. Reid, R. C., and T. K. Sherwood, *The Properties of Gases and Liquids* 2nd ed., (New York: McGraw-Hill Book Company, 1966).

17. Dobbs, (New York: A., and J. M. Cohen. "Carbon Adsorption Isotherms from Toxic Organics," U.S. EPA Report 600/8–80–023 (1980).

18. Radke, C. J., and J. M. Prausnitz. "Thermodynamics of Multi-Solute Adsorption in Dilute Liquid Solution," *Am. Inst. Chem. Eng. J.* 18:761 (1972).

19. Miller, R., and D. J. Hartman. "Feasibility Study of Granular Activated Carbon Adsorption and On-Site Regeneration," U.S. EPA 600/2-82-087A (1982).

20. Lykins, B. W. Jr., E. E. Geldreich, J. Q. Adams, J. C. Ireland, and R. M. Clark. "Granular Activated Carbon for Removing Nontrihalomethane Organics from Drinking Water," U.S. EPA Drinking Water Research Division WERL (September 1984).

21. Symons, J. M., A. A. Stevens, R. M. Clark, E. E. Geldreich, O. T. Love, Jr. "Treatment Techniques for Controlling Trihalomethanes in Drinking Water," U.S. EPA Report 600-12-81-156 (1951).

22. Perry, R. H. and C. H. Chilton, Eds. *Chemical Engineer's Handbook*, 5th ed. (New York: McGraw-Hill Book Company, 1973), pp. 16–23.

23. Traybal, R. E. *Mass Transfer Operations*, 3rd ed. (New York: McGraw-Hill Book Company, 1980), pp. 565–567.
24. Clark, R. M., J. M. Symons, and J. C. Ireland. "Evaluating Field-Scale GAC Systems for Drinking Water," *J. Environ. Eng.* 112(4):774–756 (1986).
25. Clark, R. M. "The Logistic Function," paper submitted to Xavier University, Master of Science (1964).

CHAPTER 10

Control of Trihalomethanes and Synthetic Organics

INTRODUCTION

As has been shown from the EPA field-scale studies, GAC is very effective for removing both humic material and synthetic organic chemicals. In this section a more detailed analysis of this effect is presented. Removal of humic material as characterized by TOC also has the potential for controlling trihalomethanes.

The EPA issued an amendment to the National Interim Primary Drinking Water Regulations on November 29, 1979 establishing a maximum total trihalomethane (TTHM) level in drinking water of 0.10 mg/L.[1] This amendment has caused many drinking water utilities to alter their treatment methods to reduce the concentration of TTHMs. Various treatment techniques available to the utilities have been documented.[2]

Recently, some regulatory agencies, including the EPA, have suggested that the allowable TTHM concentrations in drinking water may be lowered.[3] This is due in part to the World Health Organization's recommendation of a guideline value of 30 μg/L of chloroform based on additional health data.[4] In addition, the June 1986 Amendments to the Safe Drinking Water Act require the EPA to develop disinfection regulations that include control of disinfection by-products.[5]

These disinfection by-product regulations could also lead to a lower TTHM regulation, forcing many drinking water utilities to change disinfectants and/or consider using various treatment technologies ranging from improved conventional treatment to GAC adsorption. Some water utilities may be able to maintain TTHM concentrations below the existing promulgated regulation of 0.10 mg/L by proper conventional treatment. However, if allowable TTHM levels are reduced substantially, GAC may be an acceptable treatment alternative, especially if chlorine has to be used to maintain a disinfectant residual in the distribution system.

BACKGROUND

Proper conventional treatment (coagulation, flocculation, sedimentation, and filtration) will reduce TTHM precursors. The specific coagulation process influences the amount as well as the TTHM reactivity of the residual organic matter remaining after treatment prior to chlorination.[6] Higher-molecular-weight organics are most effectively removed during pretreatment; lower-molecular-weight organics are effectively reduced by GAC.[7,8] Jodellah and Weber[8] indicated that increased TOC removal by activated carbon treatment resulted in decreased TTHM formation in treated water.

Several investigators have identified the benefits of coagulation prior to activated carbon adsorption. Randtke and Jepsen reported significant increases in the adsorption capacity of organics after alum coagulation.[9] Lee and co-workers showed that alum coagulation enhanced both carbon adsorption capacity and the rate of uptake.[10] Semmens and co-workers observed improved GAC performance with greater levels of pretreatment.[11] Weber and Jodellah noted that alum coagulation improved overall adsorbability of TOC.[12] Pretreatment has been identified as influencing GAC adsorption and data on removal of TOC and TTHM formation potential by conventional treatment have been presented previously.[13] Conventional treatment where chlorine is used for disinfection is generally inadequate for removing TTHM precursors to levels that will be consistent with meeting current and potentially lower TTHM regulations.

GAC has been used for many years to control taste and odor and more utilities are seriously considering its use for trace organics removal. GAC may also be a viable treatment alternative for controlling TTHMs and their precursors. Data presented by Symons and co-workers showed that the time to GAC exhaustion for TTHMs (service time until effluent concentration nearly equaled influent concentration) for 11 locations ranged from 3 to > 26 weeks.[2] No definite pattern was apparent from these data to correlate EBCT or GAC influent concentration to service time. Blanck reported percent THM reductions ranging from 60% for fresh GAC to 29% after six months' operation at Davenport, Iowa.[14]

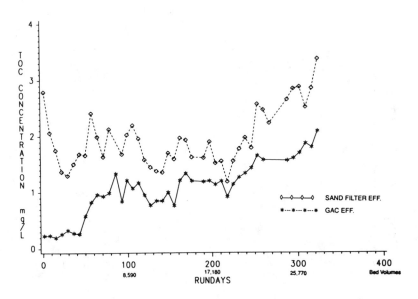

Figure 1.   TOC removal by GAC adsorption, Cincinnati, Ohio.

TOC REMOVAL BY *GAC*

TOC removal has been suggested as a means of measuring treatment performance. Although TOC is relatively easy to analyze and incorporates all organics, it does not relate to any specific regulatory requirement. In the following evaluation, however, TOC as a general surrogate parameter was used to determine the performance of GAC adsorption.

Some examples of TOC removal by GAC are presented in the following discussion. At Cincinnati, Ohio, the TOC effluent concentration at the start of one of the runs was about 0.2 mg/L (nonadsorbable fraction) and increased to about 1.1 mg/L after approximately 100 days of operation. After that point, the TOC tracked just below the sand filter effluent for the duration of the 320 days of operation (Figure 1). With Manchester, New Hampshire, the nonadsorbable TOC fraction at the start of one evaluation was about 0.5 mg/L. The GAC effluent concentration increased until about 35 days of operation before tracking just below the sand filter effluent for the remainder of the 130-day run (Figure 2).

Evaluation of GAC for TOC removal at Cincinnati and Manchester demonstrated typical or semitypical breakthrough curves. At Jefferson Parish, Louisiana, however, the full-scale GAC adsorber seemed to remove the TOC concentration steadily for about 160 days during one study (Figure 3). The nonadsorbable TOC fraction at the beginning of this study was 0.2 mg/L. This phenomenon was not seen in pilot GAC series contactors at Jefferson Parish. Using the same source water that was used for the full-

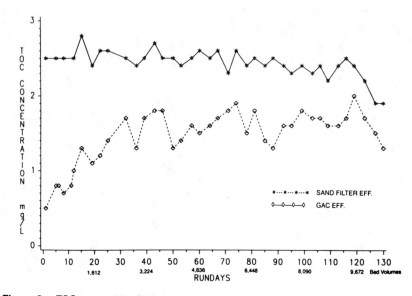

Figure 2. TOC removal by GAC adsorption, Manchester, New Hampshire.

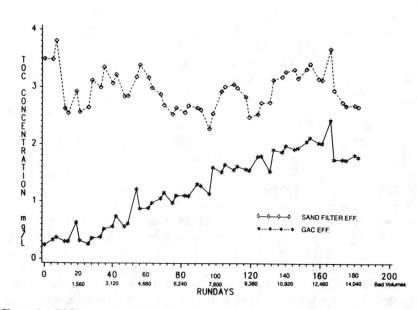

Figure 3. TOC removal by GAC adsorption, Jefferson Parish, Louisiana.

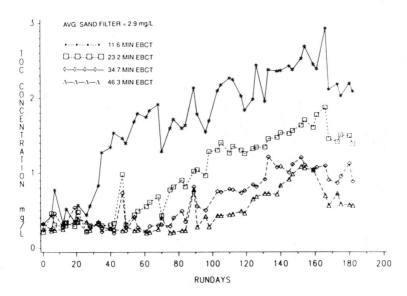

Figure 4. TOC removal through series GAC contactors, Jefferson Parish.

scale system and also experiencing a nonadsorbable fraction of 0.2 to 0.3 mg/L, the series GAC contactors showed what one might classify as typical TOC breakthrough curves, as shown in Figure 4. TOC concentrations broke through after approximately 30, 50, 80, and 110 days of operation for the 11.6-, 23.2-, 34.7-, and 46.3-min EBCTs, repectively. At Miami, Florida, the average TOC concentration of the GAC influent was higher and breakthrough occurred sooner (Figure 5).

TOX REMOVAL BY *GAC*

Total organic halide is indicative of total halogenated organics, which may be a better surrogate indicator of TTHMs than TOC. TOX removal by GAC was evaluated during field-scale evaluations at Cincinnati and Jefferson Parish. Figure 6 shows instantaneous TOX removal at Cincinnati. Less than 5 μg/L of TOX was nonadsorbable at the start of this study. A gradual increase in GAC effluent concentration was noted until about 140 days of operation. At that time, the GAC influent concentration increased with a corresponding increase in effluent concentration.

During a pilot-plant study at Jefferson Parish, chlorine was added to one of the treatment systems. The TOX influent concentration in this system varied from 120 μg/L to 340 μg/L over 370 days of operation. Figure 7 shows that the nonadsorbable TOX was less than 5 μg/L (similar to Cincinnati as noted above) with removal of the adsorbable TOX continuing throughout the 375 days of operation.

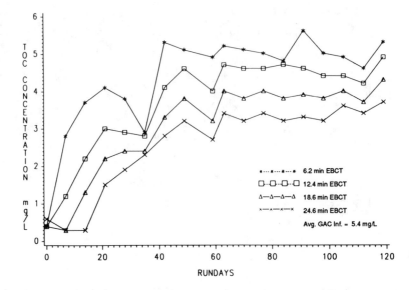

Figure 5.   TOC removal by GAC, Miami, Florida.

TTHM REMOVAL BY *GAC*

Removal of instantaneous trihalomethanes and their precursors by GAC to meet a TTHM regulation was also evaluated. Since the utilities studied used various disinfectants that affected the trihalomethane concentrations, a common parameter was needed for comparison among utilities. Terminal trihalomethane (term THM) was selected because it represented the maximum THM (instantaneous THM plus THMs formed) in the distribution system at a given time.

During the Cincinnati study, clearwell samples were stored at ambient temperature with an additional 2.5 mg/L of free chlorine residual to match conditions in the actual distributed water.[15] Figure 8 shows that term THM values were comparable for the distribution system samples and the three-day stored samples. At the other research sites, ambient conditions were maintained but often chlorine doses were used that were much greater than those used in actual treatment. This may appear to be atypical, but in most cases only a slight influence on trihalomethane formation rate or yield occurs when the free chlorine dose is increased beyond the demand.[2] Therefore, it is important to maintain a free chlorine residual in the distribution system as well as a simulated system when evaluating THM concentrations.

Some water utilities are able to maintain their TTHM concentrations below the existing promulgated regulation of 0.10 mg/L (100 μg/L) by proper conventional treatment. If the regulation is lowered substantially, however, other treatment alternatives will be required, and GAC may be an

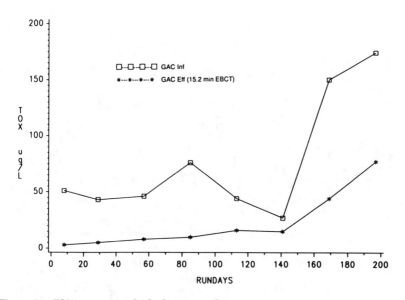

Figure 6.   TOX removal by GAC, Cincinnati, Ohio.

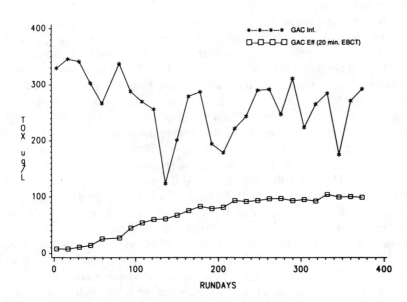

Figure 7.   TOX removal by GAC, Jefferson Parish, Louisiana.

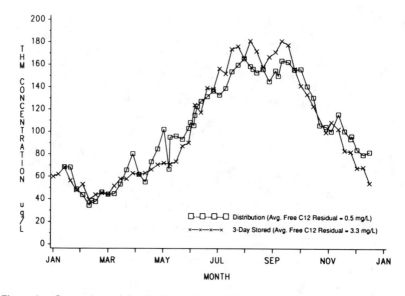

Figure 8.   Comparison of distribution systems vs three-day simulated distribution, Cincinnati, Ohio.

alternative worth evaluating. The length of time that GAC can remove trihalomethanes and their precursors to meet a possible regulation of 10 μg/ L, 25 μg/L, 50 μg/L, or 100 μg/L will determine its efficacy as a viable treatment option.

Using the same locations and runs for examples of TTHM and precursor removal by GAC as were used in the TOC examples, breakthrough curves show the same general shape. For instance, at Cincinnati, the nonadsorbable three-day term THM was about 3 μg/L at the start of an adsorption study and breakthrough occurred after about 50 days of operation. From about Day 110, the three-day term THM effluent was approximately the same increment below the sand filter effluent throughout the 320-day study (Figure 9). For Manchester, New Hampshire, the nonadsorbable three-day term THM concentration was about 10 μg/L. This concentration increased to about 45 μg/L after 40 days of operation, and then tracked below the sand filter effluent for the rest of the 130-day run (Figure 10). The nonadsorbable five-day term THM at Jefferson Parish was about 15 μg/L; this concentration steadily increased for about 140 days of operation for one run (Figure 11). With the series contactors at Jefferson Parish, the nonadsorbable fraction was about 25 μg/L and breakthrough was noted at 40, 70, 90, and 110 days for 11.6, 23.2, 34.7, and 46.3 minutes EBCT, respectively (Figure 12). The term THM concentrations for Miami were much higher than seen for the other locations evaluated and, as with TOC, broke through quickly (Figure 13).

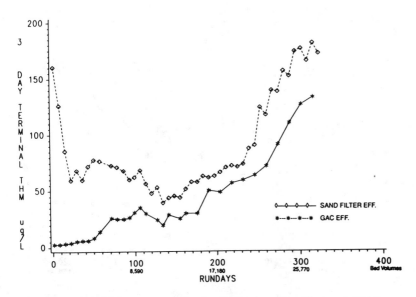

Figure 9. Terminal THM concentrations after GAC adsorption, Cincinnati, Ohio.

Since terminal THM values can simulate concentrations in the distribution system, one can estimate the length of GAC operation for meeting possible TTHM goals. Table 1 gives an indication of how long GAC can remove various concentrations of TTHMs.

As can be seen from Table 1, the deep bed GAC adsorbers that were used at Cincinnati produced the overall best terminal THM removals. For all locations evaluated, establishing a trihalomethane regulation of 10 µg/L will probably negate the use of GAC. Using GAC to meet a 25-µg/L regulation also may not be feasible in most cases. However, at the 50-µg/L THM concentration, GAC may be attractive for some surface water plants. For locations such as Miami, GAC may not be an acceptable alternative.

PREDICTING *THM* CONCENTRATIONS

Good correlations between TOC and TTHM formation potential have been presented previously.[13] Once these correlations are developed, one may be able to use surrogate parameters such as TOC or TOX as an indicator of TTHM concentration with relative analytic ease.

As presented earlier in this paper, term THM can simulate the distribution system THM concentrations. TOC correlated to three-day term THM for Cincinnati and Manchester GAC effluent shows a reasonable relationship until GAC exhaustion. A 1.0-mg/L TOC exhaustion criterion may be acceptable for THM control as well as control of other organics.[16] At 1.0-

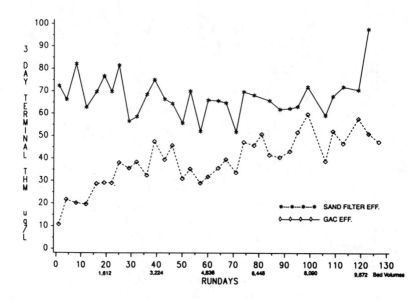

Figure 10.   Terminal THM concentrations after GAC adsorption, Manchester, New Hampshire.

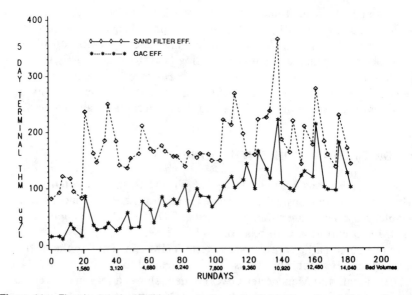

Figure 11.   Five-day terminal THM concentrations after GAC adsorption, Jefferson Parish, Louisiana.

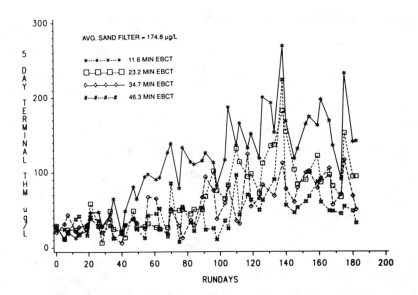

Figure 12. Five-day terminal THM for GAC series contactors, Jefferson Parish, Louisiana.

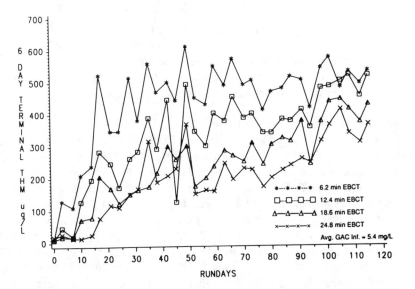

Figure 13. Six-day terminal THM for GAC series contactors, Miami, Florida.

Table 1. Length of GAC Operation Before Exceeding Possible TTHM Goals

	TTHM Goals							
	10 μg/L		25 μg/L		50 μg/L		100 μg/L	
Location	Day	Inf. (μg/L)	Day	Inf. (μg/L)	Day	Inf. (μg/L)	Day	Inf. (μg/L)
Cincinnati, Ohio (three-day term, 15.2 min EBCT)	50	75	155	45	208	70	208	150
Jefferson Parish, Louisiana (five-day term, 18.8 min EBCT)	—	—	20	80	63	170	103	220
Manchester, New Hampshire (three-day term, 23 min EBCT)	2	73	16	70	98	65	—	—
Evansville, Indiana (three-day term, 9.6 min EBCT)	—	—	6	96	56	53	—	—
Miami, Florida (six-day term, 24.8 min EBCT)	—	—	15	580	17	580	20	510

mg/L GAC exhaustion for Cincinnati, about 32 μg/L term THM was seen (Figure 14). The same GAC exhaustion criterion for Manchester gave about 25 μg/L term THM (Figure 15). Although the 1.0-mg/L TOC criterion may allow one to keep THM concentrations within acceptable levels, length of GAC operation may also be short. For instance, Manchester would only be able to operate for approximately 16 days. Conversely, with deep bed GAC adsorbers, Cincinnati could operate much longer (> 155 days).

In order to illustrate the relationship between term THM and TOC, the following equation was hypothesized for term THM breaking through GAC as related to TOC:[17]

$$\text{term THM} = A \text{ TOC}^a \tag{1}$$

where term THM = terminal trihalomethane, μg/L
    TOC = total organic carbon, mg/L
    A, a = parameters determined by regression

Regression results are shown in Table 2.

TOX was also evaluated as a possible predictor of TTHM concentrations. Instantaneous TTHM and instantaneous TOX of the GAC effluent at Cincinnati tracked each other throughout 210 days of operation (Figure 16). The instantaneous TTHM concentration was always lower than the TOX, as might be expected since the TOX includes all of the halogenated organics. Comparing these two parameters on a μmole basis did not produce a closer relationship. Instantaneous TTHM and instantaneous TOX were compared at Jefferson Parish for 375 days of operation. Up to about 90 days of operation, the instantaneous TTHM tracked below the TOX as was seen in Cincinnati (Figure 17). After that point in time, the two parameters did not correlate as well.

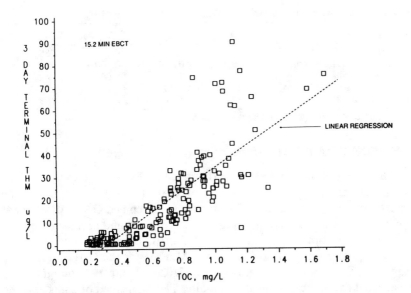

Figure 14. Correlation between three-day terminal THM and TOC for GAC effluent, Cincinnati, Ohio.

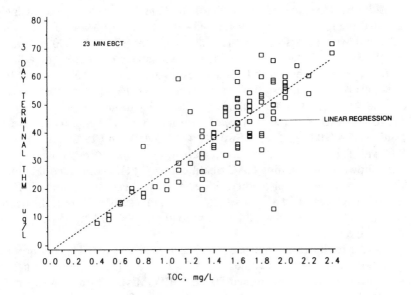

Figure 15. Correlation between three-day terminal THM and TOC for GAC effluent, Manchester, New Hampshire.

Table 2. Regression Results for Term THM vs TOC

Location	A	a	R^{2a}	N
Cincinnati, Ohio	26.5	2.002	0.814	735
Jefferson Parish, Louisiana	68.4	0.616	0.460	404
Evansville, Indiana	23.4	1.235	0.661	47
Manchester, New Hampshire	25.1	1.407	0.743	88
Miami, Florida	53.6	1.372	0.906	68

$^a R^2$ = Square of correlation coefficient (variance explained).

COST FOR REMOVING *TTHMs*

By looking at the days to GAC exhaustion in Table 1 for various THM goals, one can speculate that the use of GAC for THM control will be expensive except in the case of Cincinnati. The deep bed GAC adsorbers at Cincinnati were designed to provide efficient organic and organic precursor removal. On the other hand, at Jefferson Parish, an existing sand filter was used as a postfiltration adsorber, and at Manchester the GAC filters were designed for taste and odor control. If a GAC system can be designed and constructed for organics control, including THMs, instead of using existing structures, GAC may be cost-effective for THM removal in some locations.

Table 3 shows cost estimates for THM removal by GAC for the locations mentioned above. Actual days to exhaustion for each THM goal were determined for each location, and costs were generated from a computer program that was developed with data from GAC field operations. All five locations could remove terminal THMs to 25 μg/L at varying costs. Cincinnati had the lowest cost at 3.9 ¢/m³ (14.9 ¢/1000 gal); Manchester, Evansville, and Miami costs were in the 24.3–25.8 ¢/m³ (92–98 ¢/1000 gal) range. For the 50-μg/L THM goal, Cincinnati's estimated cost of 3.6 ¢/m³ (13.8 ¢/1000 gal) was the lowest and Miami's estimated cost of 21.9 ¢/m³ (82.9 ¢/1000 gal) was the highest. Jefferson Parish's and Evansville's costs were in the 6.6 ¢/m³ (25 ¢/1000 gal) range. Three locations had GAC influent THMs that exceeded 100 μg/L. Costs for meeting a 100-μg/L goal were 3.2 ¢/m³ (12.2 ¢/1000 gal), 5.5 ¢/m³ (21 ¢/1000 gal), and 19.0 ¢/m³ (72.1 ¢/1000 gal) for Cincinnati, Jefferson Parish, and Miami, respectively.

As shown above, adding GAC treatment for TTHM control could place a large cost burden on all of the utilities evaluated except Cincinnati. If the cost of GAC treatment at Cincinnati is 4.0 ¢/m³ (15 ¢/1000 gal), the overall increase in the water bill would amount to about 20% or $20–25 per year for an average household.

In order to illustrate the use of TOC as a predictor of TTHM concentration, Equation 1 and the values in Table 2 were utilized to predict a TOC concentration from an assumed 50-μg/L TTHM value. Table 4 shows the results of this prediction and the target level for TOC that will result in a 50-μg/L concentration for TTHMs in the distribution system. Using these

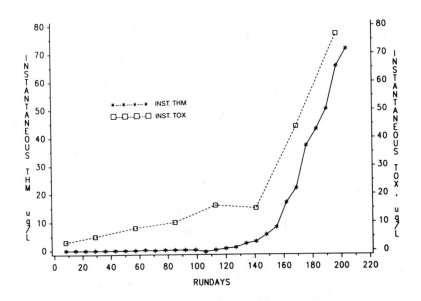

Figure 16. Comparison of GAC effluent instantaneous THM and TOX, Jefferson Parish, Louisiana.

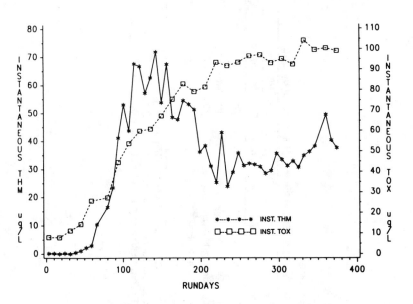

Figure 17. Comparison of GAC effluent instantaneous THM and TOX, Cincinnati, Ohio.

Table 3. Cost Estimates for THM Removal by GAC

Location	10 µg/L			25 µg/L			50 µg/L			100 µg/L		
	day[a]	Co[b]	¢/1000 gal[c]	day[a]	Co[b]	¢/1000 gal[c]	day[a]	Co[b]	¢/1000 gal[c]	day[a]	Co[b]	¢/1000 gal[c]
Cincinnati, Ohio	50	75	24.9	155	45	14.9	208	70	13.8	280	150	12.2
Jefferson Parish, Louisiana	—	—	—	20	80	57.4	63	170	25.8	103	220	21.0
Manchester, New Hampshire	2	73	56.9	16	70	98.0	98	65	41.2	—	—	—
Evansville, Indiana	—	—	—	6	9	92.1	56	53	25.7	—	—	—
Miami, Florida	—	—	—	15	580	92.7	17	580	82.9	20	510	72.1

[a]Day to GAC exhaustion for THM goal.
[b]GAC influent concentration.
[c]To convert ¢/1000 gal to ¢/m^3, multiply by 0.264.

Table 4.   TOC Levels Equivalent to 50 μg/L TTHM

Location	TTHM μg/L	TOC μg/L
Cincinnati, Ohio	50	1.4
Jefferson Parish, Louisiana	50	0.6
Evansville, Indiana	50	1.8
Manchester, New Hampshire	50	1.6
Miami, Florida	50	1.0

TOC levels for various EBCTs yields the costs shown in Figure 18. This assumption was used to present an example of cost differences among three utilities where multiple EBCTs were available. These utilities were located at Miami, Jefferson Parish, and Cincinnati.

Figure 18 shows estimated GAC treatment costs for a 1.0-mg/L TOC target. Miami's cost ranged from $0.3 ¢/m³ ($1.27 per 1000 gal) at 12.4 min EBCT to 21.0 ¢/m³ (79.7 ¢/1000 gal) for 24.8 min EBCT. Jefferson Parish's cost did not change much, ranging from 9.2 ¢/m³ (35 ¢/1000 gal) for 10.4 min EBCT to 8.2 ¢/m³ (31 ¢/1000 gal) for 31.4 min EBCT. For Cincinnati, the cost at 4.4 min EBCT was 16.1 ¢/m³ (61 ¢/1000 gal). At a more practical EBCT of 15.2 min the cost was 3.5 ¢/m³ (13.3 ¢/1000 gal), which was comparable to the cost shown in Table 2.

As stated previously, Cincinnati's GAC system design and associated treatment costs were amenable for removing THMs. Using Cincinnati as an example, what would be the cost impact of relaxing or making more strin-

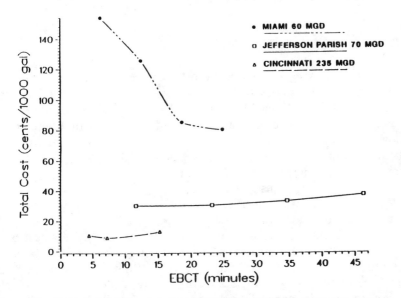

Figure 18.   Estimated GAC treatment costs for a 50-μg/L TTHM exhaustion criterion.

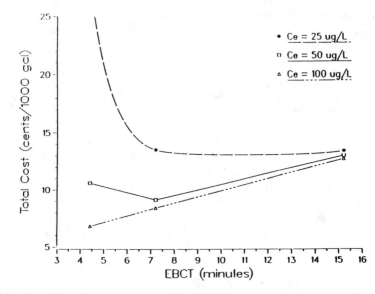

Figure 19. Estimated GAC treatment costs for various TTHM exhaustion criteria, Cincinnati, Ohio. (To convert from ¢/1000 gal to ¢/m³, multiply by 0.264.)

gent the TTHM exhaustion criterion? Figure 19 shows the cost impact. With a less stringent exhaustion criterion, there would be less reactivation required, producing a lower least-cost EBCT. The reverse would be noted with a more stringent exhaustion criterion.

REMOVAL OF SYNTHETIC ORGANIC CHEMICALS

The Safe Drinking Water Act (SDWA) requires that the EPA establish maximum contaminant levels (MCLs) for each contaminant that may have an adverse effect on the health of persons. Each MCLG is required to be set at a level so that no known or anticipated adverse effects on health occur, allowing an adequate margin of safety. The SDWA also requires that National Primary Drinking Water Regulations (NPDWR), establishing maximum contaminant levels or treatment techniques, and secondary drinking water regulations be established.[18]

Under the 1986 amendments to the SDWA, the EPA is to set maximum contaminant level goals (MCLGs), formerly called RMCLs, and NPDWR for 83 specific contaminants.[5] Of these 83 contaminants, MCLGs have been proposed for 26 synthetic organic chemicals (SOCs).[19] MCLs are to be set as close to MCLGs as is feasible. According to the SDWA amendments, "feasible" means feasible with the use of the best technology, treatment techniques, and other means under field conditions and not solely under laboratory conditions. The SDWA amendments also state that "granular activated carbon is feasible for the control of synthetic organic chemicals, and any technology, treatment technique, or other means found to be the best available for control of synthetic organic chemicals must be at least as effective as granular activated carbon."

Although regulatory pressure limits the field-scale evaluation of SOC removal approaches because of the time required, acceptable evaluations are needed to address the SDWA amendments. Some concern has been registered about EPA's approach to standard setting. The American Water Works Association (AWWA) "will support field-tested best available technology as the basis for standard setting."[20] An acceptable approach to making GAC treatability decisions for regulatory purposes may include a combination of traditional GAC evaluations and microcolumns in combination with Freundlich isotherms and modeling.

This section contains an overview of the adsorbability of SOCs by GAC. It demonstrates adsorbability by presenting data on the removal of SOCs and organic surrogates such as total organic carbon and total organic halide by field-scale contactors, pilot columns, microcolumns, and isotherms.

Synthetic Organic Regulations

Of the 83 contaminants, 43 SOCs were considered for regulation by the EPA. Inclusion of specific SOCs on the list was based on the occurrence or potential occurrence of the SOC in drinking water and the potential health effects of exposure to that SOC. Of the 43 SOCs, 26 have been proposed for regulation. Table 5 shows the proposed MCLGs for these 26 SOCs.

Other SOCs will be considered in later phases for regulation as additional health and occurrence data become available. SOCs included in this group are adipates, dalapon, dibromomethane, dinoseb, diquat, endothall, glyphosate, hexachlorocyclopentadiene, polyaromatic hydrocarbons, phthalates, picloram, 1,1,2-trichloroethane, and oxamyl.

Removal of Surrogate Parameters: Total Organic Carbon

One of the surrogate parameters often used to give an indication of GAC system operation is TOC. With this parameter, one not only is including regulated SOCs but also naturally occurring and other synthetic organic compounds. Because of its relative analytic ease, TOC is sometimes used as the limiting criterion for determining exhaustion of GAC.

Selection of an appropriate TOC concentration to use as the limiting criterion has mainly been arbitrary. In the Cincinnati study, 1.0 mg/L TOC was used.[15] By using 1.0 mg/L TOC as the limiting criterion for GAC exhaustion, one may be able to reactivate or replace GAC appropriately for SOC control. This possibility is being examined. A preliminary study using

Table 5. Synthetic Organic Chemicals and Proposed MCLGs

SOC	Proposed MCLG (mg/L)
Acrylamide	zero
Alachlor	zero
Aldicarb, aldicarb sulfoxide, and aldicarb sulfone	0.009
Carbofuran	0.036
Chlordane	zero
cis-1,2-Dichloroethene	0.07
Dibromochloropropane	zero
1,2-Dichloropropane	0.006
o-Dichlorobenzene	0.62
2,4-D	0.07
Ethylene dibromide	zero
Epichlorohydrin	zero
Ethylbenzene	0.68
Heptachlor	zero
Heptachlor epoxide	zero
Lindane	0.0002
Methoxychlor	0.34
Monochlorobenzene	0.06
Polychlorinated biphenyls	zero
Pentachlorophenol	0.22
Styrene	0.14
Toluene	2.0
2,4,5-TP	0.052
Toxaphene	zero
trans-1,2-Dichloroethylene	0.07
Xylenes	0.44

Ohio River water for empirical TOC data and SOC curves produced by model predictions based on isotherms were combined to show a possible relationship for GAC exhaustion that would allow TOC to be used as the limiting criterion. The SOC predicted curves shown in Figures 20 and 21 indicate that at 1.0 mg/L TOC exhaustion, one can replace or reactivate GAC prior to SOC exhaustion. Verification of this relationship may provide utilities that use GAC for SOC control a means of determining GAC exhaustion without analyzing for each SOC.

Removal of Synthetic Organics

Field-Scale Experience

During field-scale research many of the SOCs shown in Table 5 have been detected at various research sites. Some of these SOCs were present in the source water at very low (nanogram-per-liter) concentrations, while others exceeded state or anticipated federal MCLs. Table 6 shows the SOCs found at various locations and the range of concentration.[21-23]

Generally, the locations where groundwater was used as the source had

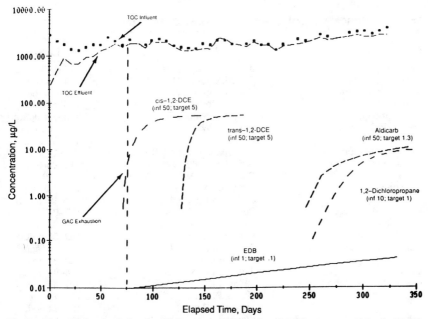

Figure 20.   GAC exhaustion for TOC with model-predicted SOC curves at 7.2 min EBCT.

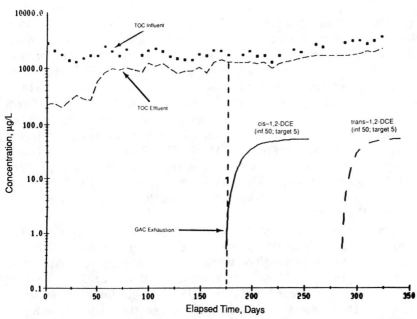

Figure 21.   GAC exhaustion for TOC with model-predicted SOC curves at 15.2 min EBCT.

Table 6.   DWRD[a] Research Locations Where SOCs Were Detected

Compound	Location	Concentration Range
Alachlor	Jefferson Parish, LA	13–593 ng/L
Aldicarb sulfone	Suffolk County, NY	14–48 µg/L
Aldicarb sulfoxide	Suffolk County, NY	12–35 µg/L
Carbofuran	Suffolk County, NY	5–13 µg/L
Chlordane	Jefferson Parish, LA	0.3–1.4 ng/L
cis-1,2-Dichloroethene	Wausau, WI	64–500 µg/L
	Miami, FL	2.4–68 µg/L
Dibromochloropropane	San Joaquin Valley, CA	0.3–13 µg/L
1,2-Dichloropropane	Suffolk County, NY	18–23 µg/L
	San Joaquin Valley, CA	ND[b]–3.0 µg/L
Ethylbenzene	Wausau, WI	ND–7.0 µg/L
	Jefferson Parish, LA	0.14–1985 ng/L
Ethylene dibromide	San Joaquin Valley, CA	ND–0.4 µg/L
Heptachlor	Jefferson Parish, LA	0.03–33 ng/L
	Passaic, NJ	ND–39 ng/L
Heptachlor epoxide	Jefferson Parish, LA	0.3–1.7 ng/L
	Passaic, NJ	ND–21 ng/L
Lindane	Jefferson Parish, LA	0.4–9.1 ng/L
Monochlorobenzene	Miami, FL	ND–5.6 µg/L
o-Dichlorobenzene	Miami, FL	ND–1.2 µg/L
	Jefferson Parish, LA	0.7–24 µg/L
Toluene	Wausau, WI	ND–120 µg/L
Xylene(s)	Wausau, WI	ND–25 µg/L
	Jefferson Parish, LA	7.0–8400 ng/L

[a]DWRD: EPA's Drinking Water Research Division.
[b]ND: Not detected.

the highest concentrations of SOCs. One example of this is Suffolk County, New York, where agricultural practices have contaminated the groundwater with aldicarb sulfoxide, aldicarb sulfone, carbofuran, and 1,2-dichloropropane.[24] Removal of these SOCs from the groundwater by GAC was investigated by using three contactors in series. One contactor (5 min EBCT) was operated until breakthrough of one of the above SOCs and then its effluent was passed through the second adsorber (total of 10 min EBCT). When breakthrough was seen in the second adsorber, its effluent was passed through the third adsorber (total of 15 min EBCT). Results from this study show that GAC effectively removed the carbofuran with no breakthrough seen through any of the adsorbers after one year of operation. The change in raw water quality for the other SOCs during the one year of operation is shown in Figure 22.

In the first GAC adsorber with a 5-min EBCT, aldicarb sulfoxide, aldicarb sulfone, and 1,2-dichloropropane were removed until about 30 days of operation (until breakthrough occurred for all three, Figure 23). At 30 days of operation, the second GAC adsorber (total of 10 min EBCT) was placed on-line. With the second GAC adsorber, 1,2-dichloropropane broke through first, followed by aldicarb sulfone and then aldicarb sulfoxide (Figure 24). The third GAC adsorber (total of 15 min EBCT) was placed in

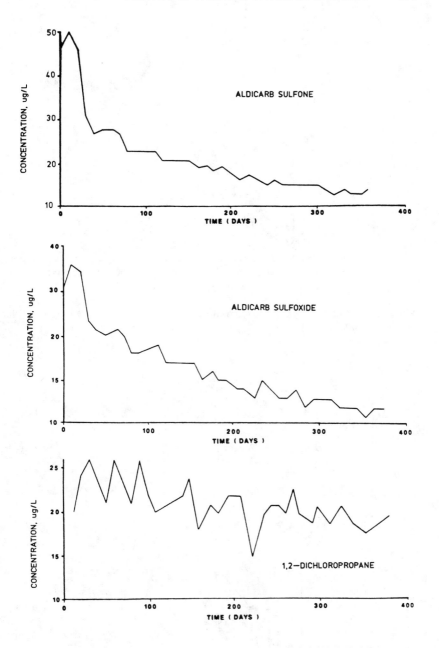

Figure 22.   Raw water concentrations for selected SOCs, Suffolk County, New York.

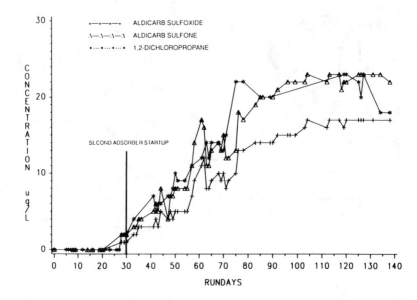

Figure 23. Organic contaminant removal by first GAC adsorber, Suffolk County, New York.

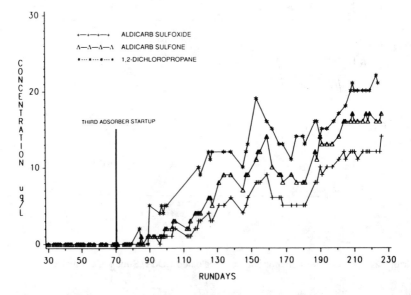

Figure 24. Organic contaminant removal by second GAC adsorber, Suffolk County, New York.

service after 70 days of operation. Again 1,2-dichloropropane broke through first (about 150 days of operation), followed by aldicarb sulfone and aldicarb sulfoxide (Figure 25). The advantage of operating the GAC adsorbers in the foregoing manner was to eliminate or reduce the effects of preadsorbed organics on GAC performance by withholding a contactor from service until breakthrough occurred in the contactor upstream. This effect has been shown to be substantial with preadsorbed TOC.[25]

Another location that uses groundwater is Wausau, Wisconsin. At Wausau, 2-in.-diameter pilot columns in series were used to determine the effects of EBCT for removing SOCs in the source water. EBCTs ranged from 1 to 30 min. The SOCs present in Wausau's water that appear in Table 6 were *cis*-1,2-dichloroethene, toluene, ethylbenzene, and the xylenes (*o,p,m*). *cis*-1,2-Dichloroethene was the only SOC that broke through the GAC above 10 min EBCT. As shown in Figure 26, *cis*-1,2-dichloroethene broke through at about Day 65, 160, and 290 for 10, 20, and 30 min EBCT, respectively. Breakthroughs did occur for the other SOCs at Wausau at shorter EBCTs. Figures 27–30 show the breakthrough profiles for these SOCs at 1-, 3-, and 5-min EBCTs.

cis-1,2-Dichloroethene occurred in the highest concentration during a research effort at Miami, Florida with treated water concentrations averaging about 21 μg/L. Removal of the *cis*-1,2-dichloroethene through GAC at various EBCTs is shown in Figure 31. Breakthrough occurred about Day 15, 80, and 120 for 6.2, 12.4, and 18.6 min, respectively.

Although the surface water treatment plants using river water as their source generally had more SOCs at lower concentrations, this does not diminish the significance of removing these compounds from the finished water delivered to the consumer. Many regulators and water utility managers are concerned not only about the acute effect of organic contamination, but also the cumulative effect of ingesting many organics at small concentrations over long periods of time.

At Jefferson Parish, several chlorinated hydrocarbon insecticides (CHIs) were detected at concentrations averaging less than 10 ng/L. Many of these CHIs are listed in Table 7. The total CHI concentration prior to GAC adsorption ranged from 18 to 88 ng/L with an annual average of 36 ng/L. The CHIs were effectively removed by GAC during one year of evaluation using a 20-min EBCT (Figure 32). Another CHI on the SOC list detected at higher concentrations was alachlor. Concentrations of alachlor ranged from 13 to 593 ng/L during the one-year evaluation with an average of 127 ng/L. Removal of alachlor by GAC with 20 min EBCT was very effective during this study (Figure 33).

Another utility using river water as its source that has detected many organics and seen removal through GAC is Cincinnati. Figure 34 shows a gas chromatographic (GC) profile of these organics in the conventional plant effluent and removal through GAC after 15 min EBCT. Many of the

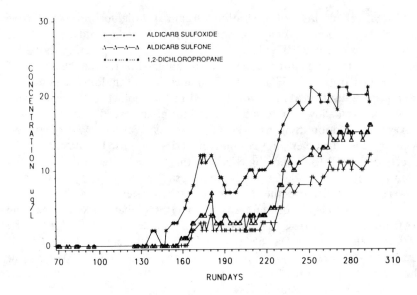

Figure 25. Organic contaminant removal by third GAC adsorber, Suffolk County, New York.

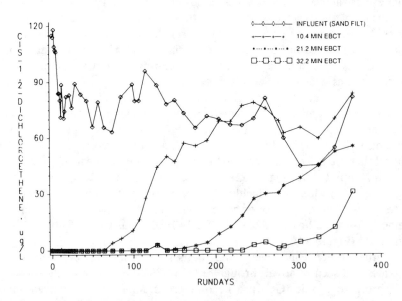

Figure 26. Removal of *cis*-1,2-dichloroethane by GAC, Wausau, Wisconsin.

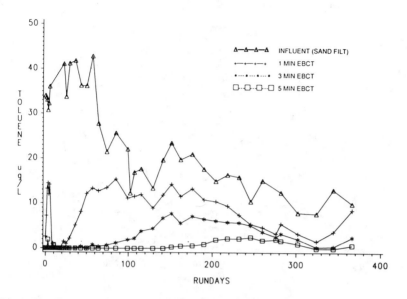

Figure 27.   Removal of toluene by GAC, Wausau, Wisconsin.

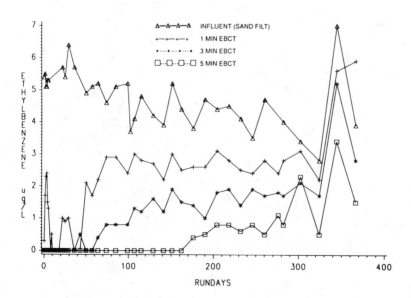

Figure 28.   Removal of ethylbenzene by GAC, Wausau, Wisconsin.

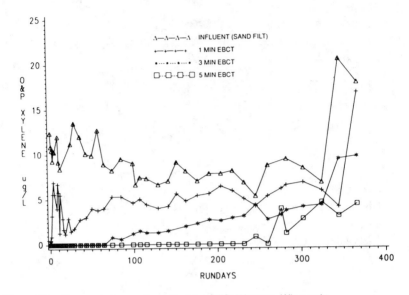

Figure 29.   Removal of *o*- and *p*-xylene by GAC, Wausau, Wisconsin.

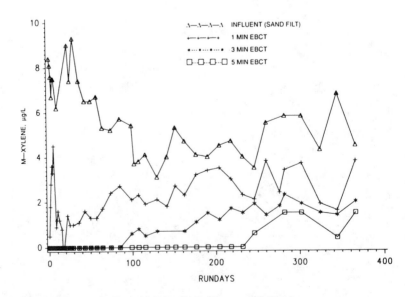

Figure 30.   Removal of *m*-xylene by GAC, Wausau, Wisconsin.

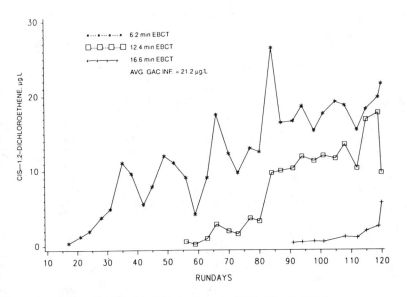

Figure 31. Removal of *cis*-1,2-dichloroethene by GAC, Miami, Florida.

peaks controlled by GAC represent various organic compounds, including SOCs. The GC profile, like TOC and TOX, demonstrates GAC's ability to perform as a broad spectrum adsorbent.

Sometimes the quality of source water used by drinking water utilities is affected by several external situations. One of these can be the proximity of landfill containment of hazardous wastes. Many of these sites are now required to provide treatment technologies to produce acceptable effluents that will not alter existing water quality. One of the most widely used treatment technologies to remove dissolved organics from aqueous wastes is GAC. Table 8 shows the effectiveness of GAC for removing several SOCs of immediate interest to EPA and water utilities that were detected in hazardous waste effluents at several locations.[26]

Prediction of GAC Adsorption

With the many SOCs scheduled for regulation and several more to be considered later, water utilities need to consider cost-effective ways of evaluating GAC for their situation. One method is to use pilot systems for predicting full-scale GAC performance. This was shown to be a viable method during studies at Cincinnati and Jefferson Parish.

Using 3-in.-diameter columns with 30 in. of GAC at Cincinnati, with approximately the same hydraulic loading and EBCT as the full-scale system, comparable TOC removals were noted. The pilot system predicted the full-scale adsorber performance within 10% in most cases for over 100 days of operation.

Table 7.   Chlorinated Hydrocarbon Insecticides Detected at Jefferson Parish, Louisiana

CHI	Frequency (%)	Detected Concentration (ng/L)		
		Min.	Max.	Avg.
Aldrin	100	0.20	0.92	0.53
α-BHC	100	1.14	6.75	2.78
β-BHC	100	0.87	3.33	1.64
γ-BHC (lindane)	100	0.40	9.06	1.39
α-Chlordane	100	0.12	0.47	0.27
γ-Chlordane	100	0.15	0.55	0.31
Oxy-chlordane	92	0.02	0.35	0.09
α-Chlordene	89	0.02	0.46	0.14
β-Chlordene	93	0.08	32.50	7.55
γ-Chlordene	96	0.06	0.77	0.43
DCPAa	100	0.17	5.42	1.42
o,p'-DDD	100	0.13	1.13	0.44
m,p'-DDD	56	0.03	1.34	0.36
p,p'-DDD	93	0.17	1.28	0.57
o,p'-DDE	96	0.08	0.83	0.30
p,p'-DDE	100	0.06	0.32	0.16
o,p'-DDT	59	0.03	7.80	0.68
p,p'-DDT	85	0.07	1.25	0.36
Dieldrin	100	1.63	5.80	2.84
Endosulfan-2	96	0.02	0.72	0.29
Endrin	100	0.25	1.58	0.67
Heptachlor	92	0.03	33.30	6.95
Heptachlor epoxide	100	0.28	1.68	0.73
Hexachlorobenzene	100	0.35	1.59	0.66
Pentachloronitrobenzene	96	0.04	0.27	0.15
Perthane	77	0.86	23.40	8.99

aDimethyl-2,3,5,6-tetrachloroterephthalate.

At Jefferson Parish, the pilot columns were also 3 in. in diameter and contained 30 in. of GAC. Hydraulic loadings for the pilot and full-scale adsorbers were comparable, but the EBCTs were somewhat different (17 and 21.2 min. for full-scale and pilot adsorbers, respectively). During two studies, the pilot adsorber removed on the average 8% and 7% more TOC than the full-scale adsorber.

Another method of predicting GAC performance is the use of microcolumns. A rapid small-scale column test (RSSCT) has been developed in an attempt to predict pilot-scale fixed bed adsorber performance.[27] An RSSCT can be conducted in a fraction of the time that is required to conduct pilot studies, with much smaller volumes of water. Using an RSSCT adsorber that was 1.1 cm in diameter by 7.7 cm in length, performance comparisons with 2-in.-diameter pilot columns were done using well water at Wausau, Wisconsin.

During one parallel study, the cis-1,2-dichloroethene average influent

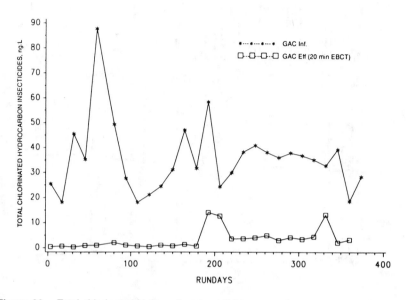

Figure 32.   Total chlorinated hydrocarbon insecticide removal.

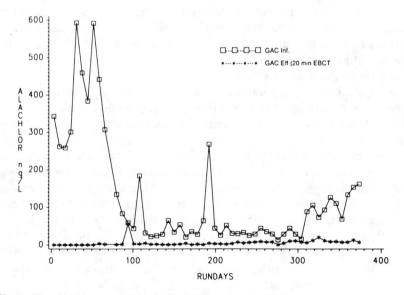

Figure 33.   Alachlor removal by GAC, Jefferson Parish, Louisiana.

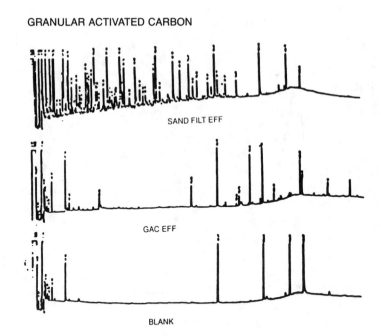

Figure 34. GC/FID capillary column profiles after 29 days of operation, Cincinnati, Ohio.

concentration was 80.9 μg/L. Several other organics were identified (Table 6) and the TOC concentration averaged 8.4 mg/L providing for competitive adsorption. For this comparison, the RSSCT satisfactorily predicted the *cis*-1,2-dichloroethene breakthrough profile when the effects of differing influent concentrations were normalized (Figure 35). Comparisons at other EBCTs were not as successful, possibly because of surface diffusivity differences between the carbon used in the pilot columns and the small-scale column. More field testing will be needed before the RSSCT can be considered a viable predicting method.

In EPA laboratories, studies are underway utilizing isotherms, microcolumns, and pilot columns in an effort to describe adsorbability and predict GAC usage rates for SOCs under regulatory consideration and in a rapid time frame.

Isotherm adsorption capacities, described by the Freundlich equation, are developed and input to the homogeneous surface diffusion model along with adsorption rate coefficients taken from the literature or experimentally determined by differential column batch reactor studies.[28] If model predictions demonstrate lower use rates and longer bed lives than those typically employed for taste and odor (T&O) control, the SOC is considered to be cost-effectively adsorbed under the assumed hydraulic and organic conditions. As an example, predicted use rates for alachlor in filtered Ohio River water in a sand-replacement contactor are given in Table 9.[16,29]

No attempt was made to verify these alachlor predictions, but verifica-

Table 8. Synthetic Organic Chemicals Removed from Hazardous Waste Streams by GAC

Compound	Location of Incident	Quantity Treated (gal)	Contact Time (min)	Influent Concentration (μg/L)	Effluent Concentration (μg/L)	Percent Removal
PCB	Seattle, WA	600,000	30–40	400	0.075	99.98
Toxaphane	The Plains, VA	250,000	26	36	1	97.22
Chlordane	Strongstown, PA	100,000	17	13	0.35	97.30
Heptachlor	Strongstown, PA	100,000	17	6.1	0.06	99.02
Pentachlorophenol	Haverford, PA	215,000	26	10,000	<0.1	99.98
Toluene	Oswego, NY	250,000	8.5	120	0.3	99.75
Xylene	Oswego, NY	250,000	8.5	140	0.1	99.92

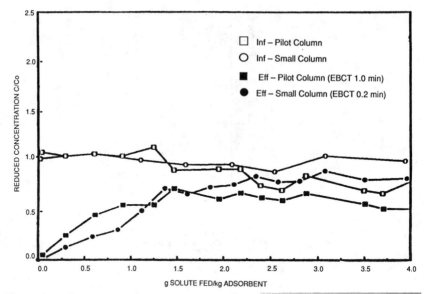

Figure 35. Comparison of *cis*-1,2-dichloroethene breakthrough profile for the small-column GAC test to the pilot-scale GAC column.

Table 9.   Predicted Full-Scale Carbon Use Rates for Alachlor[a]

Influent, μg/L	Effluent, μg/L	Use Rate, lb per 1000 gal
10	1	0.0038
	5	0.0031
20	1	0.0061
	5	0.0055
	10	0.0052
50	1	0.011
	10	0.01
	25	0.009

[a]Water: Filtered Ohio River, pH = 8, temperature = 21°C, TOC = 2.5 mg/L. Carbon: Calgon Filtrasorb 400. Contactor: 3-ft depth, 4 gpm/ft^2, 5.62-min EBCT. Freundlich isotherm: K = 328 μmol/g (L/μmol)$^{1/n}$, 1/n = 0.38. To convert from lb per 1000 gal to g/L, multiply by 0.1.

tion for the relatively weakly adsorbed SOCs is ongoing at the microcolumn and pilot-column scales. Microcolumn predictions are given in Table 10 for aldicarb in filtered Ohio River water in a postfilter contactor.[16] Like alachlor, aldicarb is relatively strongly adsorbed and pilot-column studies would require many months. Therefore, no attempt was made to verify these aldicarb predictions at the pilot scale.

Table 10.   Predicted Full-Scale Carbon Use Rates for Aldicarb[a]

Influent, μg/L	Effluent, μg/L	Use Rate, lb per 1000 gal A[b]	B[c]
50	20	0.0199	0.0217
	9	0.0203	0.0225
	1.3	0.0211	0.0237
100	20	0.0305	0.0338
	9	0.0311	0.0349
	1.3	0.0321	0.0358

[a]Water: Filtered Ohio River, pH = 8.15–8.34, temperature = 21°C, TOC = 1.9–2.2 mg/L. Carbon: Calgon Filtrasorb 400. Contactor: 8-ft depth, 4 gpm/ft², 15-min EBCT. Freundlich isotherm: K = 194 μmol/g (L/μmol)$^{1/n}$, 1/n = 0.41. Microcolumn: 0.4 cm in diameter × 4.7 cm in depth, 37.4 gpm/ft², EBCT = 0.031 min, mesh = 100 × 200. To convert from lb per 1000 gal to g/L, multiply by 0.1.
[b]Model prediction.
[c]Scaled up from microcolumn breakthrough curve.

REFERENCES

1. "Control of Trihalomethanes in Drinking Water," *Federal Register* 44 (231):68624–68707 (November 29, 1979).
2. Symons, J. M., et al. "Treatment Techniques for Controlling Trihalomethanes in Drinking Water," U.S. EPA Report 600/2-81-156 (September 1981).
3. Vogt, C. "Development of Drinking Water Regulations for Organic Contaminants in the United States," paper presented at the Second National Conference on Drinking Water Treatment for Organic Contaminants, Edmonton, Alberta, Canada, April 7–8, 1986.
4. "Guidelines for Drinking Water Quality," Work Health Organization (1984).
5. The Safe Drinking Water Act Amendments of 1986, Public Law 99–339 (1986).
6. Collins, M. R., G. L. Amy, and P. H. King. "Removal of Organic Matter in Water Treatment," *J. Environ. Eng. Div. ASCE* 111 (6) (1985).
7. Semmens, M. J., and A. B. Staples. "The Nature of Organics Removed During Treatment of Mississippi River Water," *J. Am. Water Works Assoc.* 78 (2):76–81 (1986).
8. Jodellah, A. M., and W. J. Weber. "Controlling Trihalomethane Formation Potential by Chemical Treatment and Adsorption," *J. Am. Water Works Assoc.* 77 (10):95–100 (1985).
9. Randtke, S. J., and C. P. Jepsen. "Chemical Pretreatment for Activated Carbon Adsorption," *J. Am. Water Works Assoc.* 73 (8):411–419 (1981).
10. Lee, M. C., V. L. Snoeyink, and J. C. Crittenden. "Activated Carbon Adsorption of Humic Substances," *J. Am. Water Works Assoc.* 73 (8):440–446 (1981).
11. Semmens, M. J., A. B. Staples, G. Hohenstein, and G. E. Norgaard.

"Influence of Coagulation on Removal of Organics by Granular Activated Carbon," *J. Am. Water Works Assoc.* 78 (8):80–84 (1986).

12. Weber, W. J., and A. M. Jodellah. "Removing Humic Substances by Chemical Treatment and Adsorption," *J. Am. Water Works Assoc.* 7 (4):132–137 (1985).

13. Lykins, Jr., B. W., and R. M. Clark. "Removal of Humic Material by Conventional Treatment and Carbon," paper presented at the American Chemical Society Meeting, Denver, CO, April 5–10, 1987.

14. Blanck, C. A. "Trihalomethane Reduction in Operating Water Treatment Plants," *J. Am. Water Works Assoc.* 71 (9):525–528 (1979).

15. Miller, R., and D. J. Hartman. "Feasibility Study of Granular Activated Carbon and On-Site Regeneration," U.S. EPA Report 600/S2-82-087 (1982).

16. Lykins, Jr., B. W., R. M. Clark, and R. M. Miltner. "Synthetic Organic Chemical Removal by Granular Activated Carbon," paper presented at the Annual American Water Works Association Conference, Kansas City, MO, 1987.

17. Lykins, Jr., B. W., R. M. Clark, and J. Q. Adams. "Granular Activated Carbon for Controlling THMs," *J. Am. Water Works Assoc.* 80 (5):85–92 (1988).

18. Safe Drinking Water Act, Public Law 93-523 (1974).

19. "Synthetic Organic Chemicals, Inorganic Chemicals, and Microorganisms: Proposed Rule," National Primary Drinking Water Regulations, *Federal Register*, 40 CFR Part 141 (November 13, 1985), pp. 46936–47002.

20. "AWWA Testimony Before Senate Subcommittee on SDWA Amendments," Update (May 1985).

21. Koffskey, W. E., N. V. Brodtmann, and B. W. Lykins, Jr. "Organic Contaminant Removal in Lower Mississippi River Drinking Water by Granular Activated Carbon Adsorption," U.S. EPA Report 600/S2-83-032 (1983).

22. Wood, R. R., et al. "Removing Potential Organic Carcinogens and Precursors from Drinking Water," U.S. EPA Report 600/2-80-131a (1980).

23. Lykins, Jr., B. W., W. E. Koffskey, and R. G. Miller. "Chemical Products and Toxicologic Effects of Disinfection," *J. Am. Water Works Assoc.* 78 (11) (1986).

24. Baier, J. H. "GAC and RO Treatment for the Removal of Organic Contaminants from Groundwater," U.S. EPA/AWWARF Seminar, Cincinnati, OH, March 24–26, 1987.

25. Sontheimer, H., and G. Zimmer. "Activated Carbon Adsorption of Organic Pollutants in the Presence of Humic Substances," paper presented at the Environmental Chemistry Division, American Chemical Society 193rd National Meeting, Denver, CO, April 5–10, 1987.

26. "Alternatives to Hazardous Waste Landfills," U.S. EPA Report 600/8-86/017 (1986).

27. Crittenden, J. C., et al. "Design of Rapid Fixed-Bed Adsorption Tests

for Nonconstant Diffusivities," *J. Environ. Eng. Div. ASCE* 113 (2):243–259 (1987).
28. Hand, D. W., et al. "Simplified Models for the Design of Fixed-Bed Adsorption Systems," *J. Environ. Eng. Div. ASCE* 110 (2) (1984).
29. Miltner, R. J., et al. "Removal of Alachlor from Drinking Water," National Conference on Environmental Engineering, ASCE, Orlando, FL, July 1987.

CHAPTER 11

Cost Analysis for GAC

INTRODUCTION

In addition to acquiring operating and performance information, a major objective of the field-scale projects was to collect cost data.

The field projects at Cincinnati, Jefferson Parish, and Manchester were operated over a sufficient period of time to acquire very useful cost information. The information acquired for each project is summarized here. In addition, cost estimating equations, developed from these data as well as other sources, are presented.

CINCINNATI WATER WORKS COST DATA

The cost analysis presented here is based on operation of the filter adsorbers, adsorbers, and fluid bed furnace at the Cincinnati Water Works (CWW) for the years 1979 and 1980.[1] Costs were heavily impacted by the various design parameters and O&M factors associated with the system. Table 1 summarizes the amount of water treated by the GAC filters and contactors and the quantity of carbon reactivated during the period on which the cost analysis was based.

Table 2 summarizes the various operation and maintenance factors associated with the CWW GAC system. Many of the factors varied widely over the periods during which the system was evaluated. In those cases a range of values is presented.

Table 3 presents the detailed capital costs incurred during the research project for equipment and facilities at the CWW. The costs are categorized

Table 1. System Design Parameters

Category	Item	Values	
System water throughput	GAC filters	2380 mil gal/year	(9,008 mil L/year)
	GAC contactors	1200 mil gal/year	(4,542 mil L/year)
	Total flow	3580 mil gal/year	(13,550 mil L/year)
Reactivation requirements	GAC filters	1,023,000 lb/year	(464,442 kg/year)
	GAC contactors	753,600 lb/year	(342,134 kg/year)
	Total GAC	1,776,600 lb/year	(806,576 kg/year)

into building, contactor, filter modification, and reactivation system costs. Representative operation and maintenance costs and annualized capital costs associated with the CWW GAC system are summarized in Table 4. The costs for both contactor and GAC filters are based on add-on costs to the utility. Very small capital and electrical operating costs were incurred for the GAC filters. The most significant O&M costs for both the filter and contactor systems were the total labor expense, the cost of makeup GAC, and the cost of fuel.

The GAC filter adsorber system experienced an average GAC loss of 18.5% by volume (bed-to-bed, which includes reactivation and handling). The adsorber system experienced an average carbon loss of 15.3% by volume (contactor-to-contactor). The higher losses from GAC filter adsorbers were a result of longer transfer piping runs, shovel handling of carbon, and sand separator inefficiency. Attempts were made to measure the reactivation carbon losses from the spent storage tank to the reactivated storage tank. However, the results were inconsistent and not reflective of bed-to-bed losses. Later in the study, transport losses for the adsorber system were determined to be approximately 3 to 4% of total losses.

The backwash criterion of 20 psig (138 KPa) head loss was never exceeded in the contactors during the study period. Therefore, the contactor system pumping electric costs consisted primarily of influent pumping. The contactor system capital and fluid bed reactivation capital represented very signifi-

Table 2. O&M Factors (1979–1980)

Item	Range of Values	
Power	2.19–2.91 ¢/kWh	
Natural gas	31.0–33.8 ¢/ccf	(11.1–11.9 ¢/m³)
Water	42.2 ¢/1000 gal	(11.1 ¢/m³)
GAC	49–53 ¢/lb	($1.08–1.17/kg)
Contactor and furnace labor	$8/hr	
General labor	$6.50/hr	
Maintenance labor	$7–9/hr	
Natural gas fuel value	1000 Btu/ft³	(35,300 Btu/m³)
Reactivation fuel use	6000 Btu/lb GAC	(13,200 Btu/kg GAC)

Table 3.   Cincinnati Water Works Detailed Capital Costs (1979–80)

Building	
Foundation and tunnel	$ 70,700
Building heat, ventilation, and light	175,190
Floor and drain system	52,100
Potable water piping	3,080
Final connections	8,160
Miscellaneous	4,210
Engineering, design	44,520
Engineering, resident	20,970
Total building	$ 378,940
Contactors	
Pressure vessels (4)	355,550
Sample troughs	24,310
Influent/effluent piping	156,380
Process pumps	37,370
Backwash piping	77,490
GAC transport piping	60,100
Compressed air equipment	38,130
Switchgear	53,500
Painting	12,100
Engineering, design	82,920
Engineering, resident	16,540
GAC installation labor	920
Initial GAC	85,290
Total contactors	$1,000,600
GAC Filters (2)	
Sand removal labor	2,000
GAC installation labor	1,400
Sample units	11,190
Turbidimeters	2,030
Initial GAC	96,250
Total GAC filters	$ 112,870
Reactivator	
Furnace	248,230
Instrumentation	224,500
Storage tanks (2)	53,650
Steam generator and piping	14,900
Sand separator	13,850
Motor control center and wiring	17,010
Water piping	9,810
Drain piping	7,220
GAC transport piping	38,900
Engineering, design	84,780
Engineering, resident	7,410
Final connections	38,630
Total reactivator	$ 758,890
TOTAL CAPITAL COST	$2,251,300

Table 4.   CWW 1979–1980 Representative Annual O&M and Capital Costs ($/Year)

Item	GAC Filters 10 mgd (0.44 m³/s)	Contactors 4 mgd (0.17 m³/s)	Reactivation 500 lb/hr (227 kg/hr)
GAC transport water	1,380	980	—
Process water	—	—	14,800
Pumping electric	—	12,670	—
Building electric	—	3,810	3,810
Furnace electric	—	—	4,440
Natural gas	—	—	35,520
Transport labor and GAC process	57,400	40,960	—
Furnace labor	—	—	12,500
Maintenance labor	—	—	66,850
Maintenance material	—	—	1,540
GAC makeup, filters	—	—	97,400
GAC makeup, contactor	—	—	53,920
Amortized capital	13,260	202,460	120,650
	72,040	260,880	411,430
Allocated reactivation cost	247,176	164,254	
Total ($/year)	319,216	425,134	

cant costs in the system. Maintenance costs were practically nonexistent during the research project because most of the equipment and facilities were new and covered under warranties. This is not representative of continued real-life operation and must be included in a cost estimate for complete plant conversion.

In comparing the costs for GAC filter adsorbers and contactors, the most significant difference is found in the capital costs. In the research project, no structural changes were made to convert the filter beds. Therefore, the GAC filter adsorber system required very little capital investment. However, the contactors required completely new construction, resulting in significant capital expenditures.

An examination of the onsite fluid bed reactivation expenditures shows that the capital cost represented about 29% of the total reactivation system cost and O&M represented about 71%. The total unit cost for the 12,000-lb/day (5,448-kg/day) fluid bed reactivation system was 23.16 ¢/lb (51.0 ¢/kg), based on the amount of GAC required to be reactivated. For CWW, the unit cost for purchased virgin replacement GAC was about 50 ¢/lb (110 ¢/kg). Therefore, onsite fluid bed reactivation was very cost-effective as compared to replacement of spent carbon with purchased virgin GAC.

Prorating the GAC reactivation cost (on the basis of carbon quantity) between the two types of adsorption systems results in a total unit cost of

13.41 ¢/1000 gal (3.54 ¢/m³) for the filter adsorber system and 35.42 ¢/1000 gal (9.35 ¢/m³) for the contactor system. The large difference in cost may be attributed to the large additional capital costs associated with the GAC contactor facility versus negligible additional capital costs associated with the GAC filter adsorbers. Another major factor for the differences in cost is the economies-of-scale factor. Larger systems inherently demonstrate lower unit costs due to size alone. The GAC filter adsorber system was designed as a 10-mgd facilty (37,850 m³/day) with an EBCT of 7.5 min, while the GAC contactor system was designed as a 4-mgd facility (15,140 m³/day) with an EBCT of 15.3 min. One would expect lower unit costs for a system of larger scale and less stringent treatment criteria. Therefore, these total unit costs should not be considered indicative of the costs for a complete plant conversion to GAC treatment, and one should not conclude that a sand replacement filter adsorber system is more cost-effective than a contactor system.

The contactor facilities were built primarily for research requiring additional sampling ports, etc., and therefore adding significantly to their cost. Also, the short useful life of eight years used in determining the amortized capital resulted in inflated annual cost values. Therefore, in a complete treatment plant application of a GAC contactor system, the unit amortized capital cost is expected to be significantly less than that incurred in the research project. Application of GAC filter adsorbers in a complete plant conversion might require the addition of filters and GAC storage to compensate for the reduction in plant capacity as a result of frequent filter changeovers for reactivation. Therefore, the unit capital cost for a GAC filter adsorber system could increase significantly in a complete full-scale operation as compared to the field research project.

For one to make a more valid cost comparison between a sand replacement GAC filter adsorption system and a GAC contactor adsorption system, an equal basis must be established (i.e., common plant capacity, average treatment flow, and common treatment objective).

Estimated Scale-up Costs

Cost estimates for both sand replacement GAC filter adsorber systems and concrete gravity GAC contactor systems have been generated to examine the costs associated with a complete plant conversion to GAC treatment at CWW. Table 5 presents cost estimates for four complete GAC systems (adsorption with onsite fluid bed reactivation) at the CWW.

The cost estimates are given in 1984 dollars and calculated using information derived from the research project and cost data obtained from a previous EPA study.[2] The systems examined were based on a plant capacity of 235 mgd (889,475 m³/d) and an average water production of 140 mgd (529,900 m³/day). Two alternative sand replacement GAC filter adsorber systems are presented, each having an EBCT of 7.5 min, one with a treat-

Table 5. Cost Estimates (1984 Dollars) For Cincinnati Complete Plant Conversion To GAC Treatment

	GAC Filters		GAC Contactors	
	Treatment Goal		Treatment Goal	
	1.5 mg/L TOC	1.0 mg/L TOC	1.5 mg/L TOC	1.0 mg/L TOC
EBCT (min)	7.5	7.5	15.2	20.0
Critical summer service life (days)	31	14	66	114
Capital cost ($/year)[a]	3,607,411	5,116,213	5,280,807	5,884,324
O&M cost ($/year)[b]	4,039,311	8,672,638	4,341,719	3,304,890
Total cost ($/year)	7,646,722	13,788,851	9,622,526	9,189,214
Unit cost(¢/1000 gal)[c]	14.9	26.9	18.8	17.9

[a]Amortized at 10% for 20 years; based on 235 mgd (889,475 m³/d) capacity.
[b]Based on current average water production of 140 mgd (529,900 m³/d).
[c]To determine ¢/m³, multiply by 0.2642.

ment criterion of 1.5 mg/L TOC and the other with 1.0 mg/L TOC as a treatment criterion. Two alternative concrete gravity GAC contactor systems are presented for a treatment criterion of 1.0 mg/L TOC, one using an EBCT of 15.2 min and the other an EBCT of 20 min. The cost tradeoffs between the GAC filter adsorber alternative with a treatment criterion of 1.0 mg/L TOC and the two GAC contactor alternatives (both with the 1.0-mg/L TOC criterion) can be evaluated, because all three alternative systems are compared using a common treatment objective.

In examining the capital costs for the four systems, the GAC contactors realize slightly higher costs, as one would expect with increasing depth and volume of GAC (increasing EBCT). In examining the O&M costs, the contactor system with the longest EBCT (20 min) incurs the lowest O&M cost, as it has the longest GAC service life (given the 1.0-mg/L TOC goal) and requires less frequent carbon reactivation. The other contactor system alternative (EBCT = 15.2 min) has an O&M cost of about one-half the O&M cost of the GAC filter adsorber system (EBCT = 7.5 min) given the same treatment goal (1.0 mg/L TOC). The shorter EBCT in the filter adsorber system produces a much shorter GAC service life and thus results in more frequent reactivations. The GAC filter adsorber system alternative with the 1.5-mg/L TOC goal incurs a significantly lower O&M cost than the same system with the 1.0-mg/L TOC goal because the relaxed treatment goal increases GAC service life and results in less frequent reactivations.

Given the 1.0-mg/L TOC treatment goal, the GAC contactor system with the EBCT of 20 min exhibits the lowest total unit cost at 17.9 ¢/1000 gal (4.73 ¢/m³). Even though this alternative incurs the highest capital cost, the GAC service life is long enough to provide significantly lower O&M costs than the other alternatives for the same treatment goal, resulting in the lowest total system cost. The GAC contactor system with the EBCT of 15.2 min exhibits the next lowest unit cost at 18.8 ¢/1000 gal (4.96 ¢/m³) for this treatment goal. Because of the more frequent GAC replacement, the GAC filter adsorber system given the 1.0-mg/L TOC goal exhibits the highest unit cost at 26.9 ¢/1000 gal (7.10 ¢/m³), a 50% higher cost than the best contactor system. The costs shown in Table 5 indicate that in the case of the CWW for a given treatment goal (TOC = 1.0 mg/L) concrete gravity GAC contactors would be more cost-effective than sand replacement GAC filter adsorbers at the stated EBCTs. Also, as expected, lower costs are incurred to meet a less stringent treatment goal (TOC = 1.5 mg/L vs TOC = 1.0 mg/L) because of the longer GAC service life. In terms of the cost to the consumer, it is estimated that a complete GAC treatment system of this size will cost a four-member family at 265 L per capita per day (70 gal per capita per day [gpcd]), about an additional $20/year.

Table 6.   Capital Costs (1980) for Carbon Reactivation System at Manchester, New Hampshire

Item	Total Cost ($)
Carbon reactivation building	226,900
Fluid bed system—500-lb/hr (226.8-kg/hr) capacity	528,600a
Carbon transport system, 4 filters	120,000a
Analytical equipment	38,400
Analytical supplies (glassware, lab, gases, reagents, etc.)	3,700
Contractual services (architectural and engineering fees for building design)	25,800
Personnel cost (MWW)	22,300b
Virgin GAC for test filter	49,500
Material testing services	7,100
TOTAL COST	$1,022,300

aItem included installation and equipment costs.
bItem does not include overhead costs.

MANCHESTER *GAC* SYSTEM

Costs for Onsite Reactivation

About 3.6 million lb (1.63 million kg) of the Manchester Water Works' (MWW's) carbon was reactivated during a two-year onsite reactivation research project.[3] Cost data were collected over the two-year period, which sometimes included intermittent operation. Rather than develop a cost analysis based on intermittent operation, a more representative analysis was made using the data from a period of steady-state operation. The operational period examined was June 1980 through March 1981, following improvements made to the reactivation system. The reactivator uptime operating factor for this period was about 70%. During this 10-month period selected for the cost analysis, MWW reactivated 1,857,176 lb (843,158 kg) of carbon.

The capital costs incurred in the research project for the MWW reactivation system are given in Table 6. The initial capital costs shown in Table 6 were depreciated based on an assumed straight-line depreciation life of 15 years for equipment and 35 years for the building. The capital expenditures associated with the analytical equipment and supplies (Table 7) were not included in the depreciation cost because they are primarily related to research purposes only and not representative of normal operations. Also, the carbon transport system capital expense was not included in the depreci-

Table 7. Manchester Water Works Reactivation Costs,[a] June 1980 Through March 1981

Cost Item	Total Cost	Total Units	Unit Rates
A) Makeup carbon	$131,349	213,575 lbs (96,876 kg)	$0.615/lb ($1.35/kg)
B) Labor for reactivation	36,029	4,922 man-hours	$7.32/hr
C) Labor for transportation	5,200	710.4 man-hours	$7.32/hr
D) Labor for lab	929	122.9 man-hours	$7.56/hr
E) Labor for administration	24,515	2,074 man-hours	$11.82/hr
F) Parts and service calls	49,587	N/A	N/A
G) Fuel oil	41,415	40,015 gal (151 m^3)	$1.035/gal ($274.27/m^3)
H) Electrical power	9,286	175,207 kWh	$0.053/kWh
I) Water	13,743	42,813,084 gal (162,064 m^3)	$0.321/1,000 gal ($0.848/m^3)
J) Lab supplies	929	N/A	N/A
K) Depreciation	39,587	3,570 reactivation hours	$47,505/yr for 10 months
L) Overhead	51,338	N/A	77% of all MWW labor
TOTAL	$403,907		

[a]Costs based on reactivation of 1,857,176 lb (843,158 kg) of GAC; N/A = not applicable.

Table 8. Manchester Water Works Reactivation Unit Costs,[a] June 1980 Through March 1981

Cost Item	¢/kg	¢/lb
Makeup carbon	15.59	7.07
Labor for reactivation	4.28	1.94
Labor for transportation	0.62	0.28
Labor for lab	0.11	0.05
Labor for administration	2.91	1.32
Parts and service calls	5.87	2.67
Fuel oil	4.92	2.23
Electrical power	1.10	0.50
Water	1.63	0.74
Lab supplies and outside analyses	0.11	0.05
Depreciation[b]	4.70	2.13
Overhead[b]	6.08	2.76
TOTAL COST	47.92	21.74

[a]Unit cost based on 1,856,176 lb (843,158 kg) of GAC reactivated.
[b]Depreciation includes capital debt service; overhead includes direct and indirect overhead.

ation cost because it was not used during the operational period for which the cost analyses were performed.

A semiautomatic carbon transport device was abandoned because of incompatibility with existing filter equipment, faulty new equipment, excessive maintenance, and unacceptable removal of the filter media. Therefore, a manually operated hose and eductor system was utilized. Using this method, an operator could transport approximately 2000 lb (908 kg) of GAC in about 20 to 25 min. The operator was able to remove all of the GAC from each cell section of the filter using the hose and eductor system.

Table 7 presents a detailed breakdown of the MWW onsite reactivation costs. The costs are based on actual expenses incurred during the selected 10-month period. The table displays the dollar costs, the amount of resources utilized (i.e., labor man-hours, fuel gallons, electric kilowatt-hours, etc.) and the rate at which resources were charged.

The largest single cost item in the economic analysis of onsite GAC reactivation is makeup carbon associated with the GAC losses from reactivation and onsite handling. The GAC makeup expense represents about 33% of the total reactivation system cost. During a period which included five reactivation cycles, MWW experienced total GAC losses of 11.5% by volume. Almost all of this loss is associated with the reactivation process. The total overall labor expense (including labor overhead) represents about 29% of the total system cost. The capital cost (depreciation) represents only about 10% of the total system cost.

A summary of the unit costs for the onsite reactivation operation at the MWW is presented in Table 8. The cost-effectiveness of onsite GAC reactivation is clearly demonstrated. Manchester was able to reactivate carbon during the period shown at a total unit cost of 21.74 ¢/lb (47.92 ¢/kg) versus purchasing virgin carbon at a unit cost of 61.50 ¢/lb (135.42 ¢/kg).

Subsequent Onsite Reactivation

Subsequent experience with onsite fluid bed reactivation at MWW during the years 1983 through 1987 demonstrated that it is an economical process, even when small batches of carbon are reactivated. During the months of January, February, and March of 1985, the MWW reactivated approximately 261,000 lb (118,388 kg) of spent carbon from Filter #3. The cost of reactivation was 24.9 ¢/lb (54.8 ¢/kg) compared to a virgin GAC cost of 66 ¢/lb (145 ¢/kg). Total volumetric carbon losses incurred during this period were 15%. A number of problems were experienced during this reactivation operation. Instrumentation failures, primarily associated with gas flow measurement, were a source of many operational troubles. Additional problems were associated with plugging of the dryer plate and reactivation distribution plate, and with failure in the steam supply units. In all, 12 operational shutdowns of the reactivator were experienced during this period. In spite of these problems, onsite reactivation was very cost-effective.

During the month of March 1986, the MWW reactivated approximately 260,000 lb (117,934 kg) of spent carbon from Filter #4. In contrast to the 1985 operation, these reactivation activities were conducted over a short period of time and were virtually trouble-free. The cost of reactivation was 19.5 ¢/lb (42.9 ¢/kg) versus a virgin GAC cost of 67 ¢/lb (147.6 ¢/kg). Total carbon losses incurred during this period were 17%. The overall decline in unit cost of operations in 1986 from that of 1985 (24.9 ¢/lb in 1985 to 19.5 ¢/lb in 1986) is attributed to the smoothness of operations in 1986. The process uptime factor was 99%, enabling MWW to minimize expenses for labor and utilities. The 5.4 ¢/lb (11.9 ¢/kg) reduction in reactivation unit cost from operations in 1985 to 1986 represents the significant savings achievable by minimizing shutdowns through good and efficient operations and equipment upkeep.

COMPARISON OF *CWW* AND *MWW* REACTIVATION COSTS

Because of the different system configurations, it is not possible to compare the CWW and MWW GAC adsorption systems on a unit process basis. However, the fluid bed reactivators can be compared. The fluid bed reactivators at Cincinnati and Manchester were similarly constructed and rated at the same capacity of 500 lb/hr (227 kg/hr). An examination of the onsite reactivation costs during the period 1979 to 1981 for CWW and MWW shows that they both incurred approximately the same overall cost of operation; CWW experienced a total unit cost of 23.16 ¢/lb (50.99 ¢/kg) and MWW experienced a total unit cost of 21.74 ¢/lb (47.87 ¢/kg). A direct comparison of all the cost items could not be made between the two systems because some items were defined differently or were not included in the cost analysis. The CWW presented its capital cost as an amortized value (depreciation plus interest); MWW presented capital as depreciation only. The MWW included labor for onsite GAC transport, labor for administration, and labor overhead in the cost analysis, while the CWW did not include these items. The CWW reactivation system used natural gas for fuel, while MWW used fuel oil. Also, the rates charged on various resources (e.g., dollars per man-hour, kilowatt-hour, or pound of GAC) varied between the two systems. However, the most significant factor contributing to CWW's higher reactivation total unit cost was greater GAC losses resulting in higher GAC makeup costs.

Regional Reactivation Costs

Having achieved the initial objective of evaluating the performance and cost of a full-scale onsite fluid bed GAC reactivation system at MWW in 1981, future use of the facility for providing regional reactivation service was investigated.[4] The fluid bed GAC reactivator at the MWW had a throughput capacity of approximately 3.6 million lb (1.63 million kg) of GAC per year, assuming a reasonable operating factor of about 75%. Given this, it would take less than 4 months to reactivate the entire 1 million lb (0.45 million kg) of GAC at the MWW treatment plant, in any given year. The significant excess capacity of the reactivator prompted MWW to explore the feasibility of providing offsite (regional) reactivation to other water utilities in the New England area. An initial regional reactivation effort was undertaken to evaluate the costs and performance of providing a reactivation service to other utilities.

The first step in the regional reactivation effort involved surveying water utilities in New England that use GAC and their estimated reactivation requirements. Nineteen water treatment plants were contacted and all of them expressed interest in the MWW reactivation service. After careful evaluation, a limited regional reactivation program was executed involving three water utilities: (1) Kelseytown, Connecticut Water Company, (2) Danvers, Massachusetts Water Treatment Plant, and (3) Lowell, Massachusetts Treatment Plant. Each of the three participating utilities provided approximately 40,000 lb (18,160 kg) of dry-weight carbon to be transported and reactivated. Prior to starting any reactivation for the two utilities in Massachusetts, the approval of the State Department of Environmental Quality Engineering (DEQE) was necessary. DEQE endorsed this effort and the project began in the fall of 1981.

The vehicle selected for GAC transport was an open-top trailer dump truck with a volumetric capacity of 12,000 ft³ (340 m³) which corresponded approximately with the capacity of the spent and reactivated storage tanks at MWW. The single transport truck was modified with a watertight tailgate to ensure that there would be no road spillage. The amount of GAC that could be hauled over the road was governed by state laws which stipulate that the gross weight of the vehicle and load cannot exceed 80,000 lb (36,320 kg). The empty vehicle weighed 35,000 lb (15,890 kg) and because the wet-drained GAC weighed roughly twice as much as dry carbon, two separate hauls were required to reactivate the 40,000 lb (18,160 kg) of dry carbon for each utility.

Loading and off-loading GAC from the truck was accomplished by the hose and eductor system utilized in the earlier MWW onsite reactivation operation. The time required to fill the truck was 4 to 5 hrs, while off-loading the carbon took about 1.5 to 2.5 hrs. The difference in filling time versus emptying time was because of a constant flow of carbon provided to the eductor during emptying and an intermittent flow during the filling operation. Reactivation of 20,000 lb (9,080 kg) of dry-weight carbon was accomplished in approximately 40 hrs. Carbon feed rates, temperatures, and pressures varied slightly because of the different grades of GAC and adsorbate loadings.

The first utility that participated in the regional program was the Connecticut Water Company (CWC), Kelseytown Treatment Plant, located 172 miles (277 km) from MWW. This was followed by the Danvers, Massachusetts Water Treatment Plant, and the Lowell, Massachusetts Water Treatment Plant, located 41 and 33 miles (66 and 53 km) from MWW, respectively. All three utilities were conventional water treatment plants using GAC primarily for taste and odor removal. The regional reactivation program was conducted in November and December of 1981.

A summary of the regional GAC reactivation unit costs (cents per pound) for the three participating water utilities is given in Table 9. The table presents for each utility (1) the unit total cost of regional reactivation service by MWW (including offsite GAC transport); (2) an estimated unit cost for total spent carbon replacement with virgin GAC; and (3) an estimated unit cost for onsite reactivation. The "virgin GAC replacement cost" refers to the estimated cost associated with the replacement of the entire 40,000 lb (18,160 kg) of spent carbon for each utility. The "onsite reactivation cost" refers to the estimated cost associated with a utility constructing and operating its own individual onsite reactivation facility. The onsite costs are based on the use of fluid-bed reactivation and assume the following reactivation demands for the three utilities: 96,000 lb/year (43,584 kg) for CWC; 90,000 lb/year (40,860 kg) for Danvers, Massachusetts; and 250,000 lb/year (113,500 kg) for Lowell, Massachusetts.

Table 9 clearly demonstrates the cost-effectiveness of regional reactivation as compared to virgin replacement and onsite reactivation for the three

Table 9. Summary of Manchester Water Works Regional Reactivation Costs (1981)

Participating Utility	Regional Reactivation Cost (¢/lb)[a]			Virgin GAC Replacement Cost (¢/lb)	Estimated Onsite Reactivation Cost (¢/lb)
	Transport[b]	Reactivation	Total		
Connecticut Water Co.	9.1	57.1	66.2	104.7	146.6
Danvers, Massachusetts	6.5	31.0	37.5	67.2	149.8
Lowell, Massachusetts	5.4	39.5	44.9	105.7	76.7

[a]To convert from ¢/lb to ¢/kg, divide by 0.454.
[b]Includes truck rental and use, labor, and overhead.

participating utilities.[4] The cost of regional reactivation service for the Connecticut Water Company was 66.2 ¢/lb (145.9 ¢/kg) versus complete virgin replacement at 104.7 ¢/lb (230.5 ¢/kg) and onsite reactivation at 146.6 ¢/lb (322.9 ¢/kg); for the Danvers Water Plant, regional reactivation was 37.5 ¢/lb (82.6 ¢/kg) versus replacement at 67.2 ¢/lb (148.0 ¢/kg) and onsite reactivation at 149.8 ¢/lb (329.95 ¢/kg); and for the Lowell Water Plant, regional reactivation was 44.9 ¢/lb (99.0 ¢/kg) versus replacement at 105.7 ¢/lb (232.86 ¢/kg) and onsite reactivation at 76.7 ¢/lb (168.96 ¢/kg). The onsite reactivation alternatives were not cost-effective because of the very high unit costs incurred as a result of the diseconomies of scale associated with small facilities and operations and the large amount of excess furnace capacity.

The cost of regional reactivation for the Connecticut Water Company was found to be considerably higher than the cost for the other two utilities. The main reasons for this were that CWC experienced a higher carbon loss along with a high GAC makeup purchase rate, and a significantly longer transport distance between Connecticut and Manchester contributed to higher transport equipment and labor costs. The higher carbon losses experienced with the Connecticut GAC were probably related to the type of carbon used. Connecticut Water Company used a lignite-based adsorption medium, whereas the carbon used by the Danvers and Lowell plants was coal-based. Because the lignite-based carbon is softer than coal-based, abrasion losses are suspected to be greater with this material. The total volumetric carbon loss was 23.5% for CWC, 15.3% for Lowell, and 11.5% for Danvers.

Comparison of Tables 8 and 9 reveals that the costs of various items associated with regional reactivation were higher than the costs for onsite reactivation of MWW's own carbon. Several factors contributed to this:

- Costs were high because of the inefficiency of "first-time" operation; optimum furnace conditions could not be established with a small amount of carbon.
- The unit cost for makeup carbon was higher for regional operation because greater losses were experienced by the three participating utilities and the purchase cost of the makeup carbon was higher for the small quantities of carbon involved.
- Labor costs for GAC transport were higher because of offsite GAC handling and hauling.
- Unit costs for maintenance materials were higher because total cost was based on a small amount of GAC being reactivated over a short period of time.
- Labor costs for reactivation were higher because an increased workforce was assigned to the experimental program to ensure smooth operation. (This is expected to be lower in normal regional reactivation operations.)
- Fuel, electrical power, and water costs were higher because frequent reacti-

Table 10.   Capital Cost Summary

Design and Engineering

Design, bidding, office and field engineering, and project management			$320,586
Initial cost estimate			10,446
Furnace evaluation study			10,711
Cost accounting system			1,800
		TOTAL	$343,543

Construction

Site work	$ 15,000	Piping	
Piling	60,000	Underground	$ 95,000
Concrete		Other	217,400
Pile caps	10,000	Furnace shelter	17,000
Trenches	40,000	Furnace equipment	12,000
Shelter foundation	7,000	Electrical	
Paving	15,000	Underground	25,000
Equipment		Other	45,000
GAC storage tanks	115,000	Painting	19,000
GAC contactors	225,000	Existing trailer	6,000
Pumps	12,000	Modifications	
Chlorination	15,000	Bid total	$957,400
Electrical feeder	7,000	Change orders	77,209
TOTAL	$521,000	TOTAL	$1,034,609

Furnace

Purchase price	$664,750
Spare parts	18,098
TOTAL	$682,848

Initial GAC

Contactors	$114,332
Contactor recharge	38,110
Makeup GAC	38,110
TOTAL	$190,552

TOTAL CAPITAL COSTS	$2,251,552

vator startups and shutdowns resulted in greater consumption of these resources. This could be minimized in future operations by using two trucks instead of one. A two-truck transport operation would ensure a continuous flow of carbon to the reactivator, thereby reducing the need for shutdown.

JEFFERSON PARISH

The capital costs for the 3-mgd GAC filtration and reactivation facility in 1983 dollars are presented in Table 10.[5] The total engineering cost was $343,543, including the standard design and bidding, office and field engineering, and project management, as well as original cost estimate of $10,446, a $10,711 furnace evaluation study, and a $1,800 cost accounting system. The infrared furnace was purchased directly from the manufacturer by Jefferson Parish at a cost of $664,750. Spare parts totaling $18,098 were

Table 11. GAC Transport and Reactivation Costs

GAC Transport		¢/lb	¢/kg
GAC loss		1.5	4.0
Operating labor		0.3	0.8
Water		0.2	0.5
	TOTAL	2.0	5.3
GAC Reactivation		¢/lb	¢/kg
Electricity		5.9	15.8
GAC loss		5.2	13.9
Maintenance labor		3.1	8.3
Maintenance material		1.7	4.6
Operating labor		1.4	3.8
Water		0.7	1.9
Laboratory		0.2	0.5
	TOTAL	18.2	48.8
TOTAL *GAC* TRANSPORT AND REACTIVATION COST		20.2	54.1

also purchased with the furnace. The construction cost of the 3-mgd facility was $1,034,609, which included furnace installation. The cost of initial GAC was $190,552, bringing the total capital cost to $2,251,552.

The average GAC transport and reactivation cost for the first 11 reactivations is shown in Table 11. The greatest GAC transport cost was GAC loss at 4.0 ¢/kg (1.5 ¢/lb) with operating labor and water at 0.8 and 0.5 ¢/kg (0.3 and 0.2 ¢/lb) for a total GAC transport cost of 5.3 ¢/kg (2.0 ¢/lb) of GAC reactivated. The major reactivation cost was electricity at 15.8 ¢/kg (5.9 ¢/lb) followed closely by GAC loss at 13.9 ¢/kg (5.2 ¢/lb). Maintenance was also a considerable cost with labor at 8.3 ¢/kg (3.1 ¢/lb) and materials at 4.6 ¢/kg (1.7 ¢/lb). Operating labor was 3.8 ¢/kg (1.4 ¢/lb) with water and laboratory costs at 1.9 and 0.5 ¢/kg (0.7 and 0.2 ¢/lb) for a total reactivation cost of 48.8 ¢/kg (18.2 ¢/lb). Thus, the total transport and reactivation cost is estimated to be around 54.1 ¢/kg (20.2 ¢/lb). These figures were generated using the following cost factors: makeup GAC at 201.0 ¢/kg (75 ¢/lb), electricity at 6 ¢/kWh, processed water at 6.6 ¢/m³ (25 ¢/1000 gal), maintenance labor at $9/hr plus 40% fringe, and all other labor $8/hr plus 40% fringe.

The overall O&M costs for the entire facility for a three-month reactivation cycle with a 20-min EBCT are estimated in Table 12. The cost of operation of the three 3785-m³/day (1-mgd) GAC contactors was 0.5 ¢/m³ (2.0 ¢/1000 gal) and included 0.5 ¢/m³ (2 ¢/1000 gal) for electricity, 0.2 ¢/m³ (0.6 ¢/1000 gal) for operating labor, and 0.05 ¢/m³ (0.2 ¢/1000 gal) for laboratory analyses. An additional chlorine cost of approximately 0.08 ¢/m³ (0.3 ¢/1000 gal) was not included in the cost analysis because of a recently completed pilot column study funded under a cooperative agreement. This study compared disinfected GAC influents using ozone, chlo-

Table 12.   O&M Costs for a Three-Month GAC Reactivation Cycle with 20-Min EBCT

	¢/1000 gal	¢/m³
GAC Contactor		
Electricity	2.0	0.50
Operating labor	0.6	0.20
Laboratory	0.2	0.05
	2.8	0.75
GAC Transport	1.1	0.29
GAC Reactivation	10.2	2.65
TOTAL O&M COST	14.1	3.69

rine dioxide, chlorine, and chloramines to a nondisinfected GAC influent followed by post-GAC disinfection with chloramines. Thus, based upon this study, essentially no additional chloramine would be used in the full-scale GAC facility at this location. In addition to the 0.7 ¢/m³(2.8 ¢/1000 gal) operating cost for the GAC contactors, the GAC transport and GAC reactivation costs derived from Table 11 were 0.29 and 2.65 ¢/m³ (1.1 and 10.2 ¢/1000 gal), bringing the total O&M cost to 3.69 ¢/m³ (14.1 ¢/1000 gal). For a three-month GAC reactivation cycle with a 15-min EBCT as opposed to a 20-min EBCT, the total O&M cost would be reduced to approximately 3.1 ¢/m³ (12 ¢/1000 gal).

COST MODELING

In response to the early concerns about impacts of cost on drinking water utilities, the Drinking Water Research Division of EPA initiated a study to develop standardized cost data for 99 water supply unit processes.[6] The approach was to assume a standardized flow pattern for the treatment train and then to "cost-out" the unit processes. This approach requires assumptions about such details as common wall construction and amounts of interface pipe required. After the flow pattern was established, the costs associated with specific unit processes were calculated. "As built" design and standard cost reference documents were used to calculate the amounts of excavation, framework, and materials such as concrete and steel. Information from existing plants and manufacturers was used to calculate the costs of equipment associated with a unit process. Once basic information had been calculated, capital cost curves were developed.

Derivation of Cost Curves

The construction cost of each unit process considered in this study was presented as a function of the process design parameter that was determined to be the most useful and flexible under varying conditions. Such variables as loading rate, detention time, or other conditions that can vary because of designer's preference or regulatory agency requirements were used. For example, GAC adsorber construction cost curves are presented in terms of cubic feet of contactor volume, an approach that allows various empty bed contact times to be used. Contactor operation and maintenance curves are presented in terms of square feet of surface area, because operation and maintenance requirements are more appropriately related to surface area than to contactor volume. Reactivation facility cost curves are presented in terms of square feet of hearth area for the multiple-hearth furnace, but pounds per day of reactivation capacity is used for the other reactivation technologies considered. Such an approach provides more flexibility in the use of cost curves than if the costs were related to water flow through the treatment plant. Figure 1 illustrates a typical construction cost curve for a GAC contactor.[6]

Construction Cost Components

The costs for eight principal construction components were developed and then aggregated to give the construction cost for each unit process. The eight components are: (1) excavation and sitework; (2) manufactured equipment; (3) concrete; (4) steel; (5) labor; (6) pipe and valves; (7) electrical and instrumentation; and (8) housing. These categories also provide enough detailed information to permit accurate cost updating.[6]

The construction cost curves are not the final capital cost for the unit process. The construction cost curves do not include costs for general contractor overhead and profit, administration, engineering and legal fees, fiscal determinations, and interest during construction. These items are more directly related to the total cost of the individual unit processes. Therefore, they are added following summation of the cost of the individual unit processes, if more than one unit process is required.

Operation and Maintenance Cost Components

Operation and maintenance requirements were developed for building-related energy, process energy, maintenance materials, and labor. The separate determination of building energy allows for regional variations. Energy requirements are presented in kilowatt-hours per year for electricity, standard cubic feet per year for natural gas, and gallons per year for diesel fuel. Labor is presented in hours per year, allowing local variations to be incor-

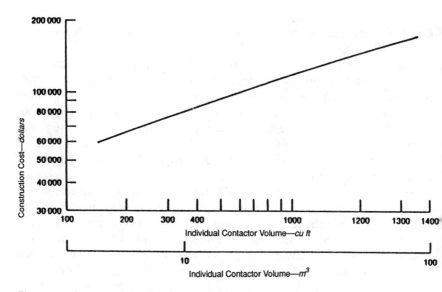

Figure 1. Construction cost for pressure carbon contactors.

porated into the operation and maintenance cost calculations. Maintenance material cost is given in dollars per year, but does not include the cost of chemicals, which must be added separately.

Figure 2 illustrates a typical operating and maintenance cost curve.[7] In order to make the cost data more flexible, the construction costs can be easily indexed using, for example, the *Engineering News Record* Construction Cost Index. Materials costs are indexed using the Producers' Price Index. Operating and maintenance input values include labor costs, electrical power cost, chemical costs, etc.

Of special interest has been the development of cost functions for GAC systems. These cost functions are based on the design studies discussed previously, but are modified by experience gained through EPA's field-scale studies.

DEVELOPMENT OF COST EQUATIONS

The cost estimating equations presented in this paper were developed in a three-step process. The first step involved developing initial mathematical functions for the basic cost components of each treatment process (i.e., for a fluid bed reactivator, the major cost components include capital, labor, electricity, fuel, and maintenance materials). The mathematical equations were developed using a nonlinear function of the form: COST $= a + bX^c Y^d$ where X and Y are independent design variables and a, b, c, and d are coefficients and exponents determined from nonlinear regres-

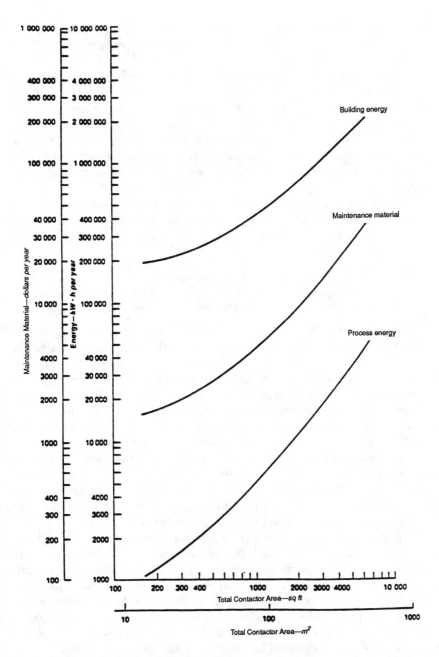

Figure 2. Operation and maintenance requirements for pressure carbon contactors, building energy, and maintenance material process energy.

sion analysis. The design and cost data used to derive the initial equations were obtained from a previous EPA cost estimating study (Culp/Wesner/Culp).[2]

The second step in the development of the equations was to compare cost estimates computed from the initial functions with actual cost data obtained from several operating systems. Cost comparisons were made using data obtained from Cincinnati Water Works, Manchester Water Works, Passaic Valley, New Jersey, and Water Factory #21.[8-11] The third step in the equation development process was to make necessary modifications to the initial cost equations as determined from the comparison of estimates with actual data.

Based on comparisons of cost data obtained from actual operating experience and estimates generated from the equations, most of the cost functions did not require major modifications. The cost equations that required some modification were:

1. capital cost for steel and concrete GAC contactors
2. process electrical energy for fluid bed and multihearth reactivation
3. O&M labor for fluid bed, infrared, and multihearth reactivation

The capital cost estimates associated with GAC contactors were low compared to actual construction data. Not all of the construction-related elements included in the actual operating system were included in the derivation of the original cost estimates on which the equations were based. Therefore, the steel and concrete GAC contactor capital cost functions were adjusted by an appropriate factor to account for these differences.

The process electrical energy estimates associated with fluid bed and multihearth reactivation were lower than actual experience. This occurred partly because the estimates did not include consumption of electrical energy during reactivator startup and shutdown. Building-related electrical energy was not taken into account for fluid bed reactivation. Therefore, these equations were also adjusted by an appropriate factor.

The reactivation O&M labor estimates were extremely low (i.e., a factor of 10) compared to actual full-scale reactivation experience. The original labor estimating data on which the initial equations were based were primarily obtained from manufacturers' estimates of reactivation requirements. Actual experience showed that even though the equipment was automated, an operator was required continuously while the reactivator was in operation. The O&M reactivation labor functions were redeveloped based on actual operating experience. The new labor functions were developed on the premise that there would be a minimum number of man-hours required to operate a reactivator regardless of its size, and that additional operation and maintenance man-hours required exhibiting economies of scale. If multiple furnaces were operated at a plant then an additional operator would be required for each furnace. Therefore, no economies of scale were assumed to exist with respect to the number of furnaces at a plant. The CWW and MWW reactivation labor data were used to calibrate the new functions, and the Culp/Wesner/Culp data were used to determine the appropriate economies-of-scale factors.

COST FUNCTION DEVELOPMENT

The cost items included in the equation development for GAC contactors, reactivation, and offsite GAC transport are listed below:

• GAC contactor capital cost
  (a) construction cost of contactor system
  (b) virgin GAC storage cost
  (c) process pumping cost (for pressure contactors)
  (d) capital overhead cost plus engineering fee
  (e) initial GAC load

• GAC contactor O&M cost
  (a) process electric cost
  (b) O&M labor cost
  (c) maintenance material cost
  (d) process pumping electric cost (pressure contactors)
  (e) process pumping labor cost (pressure contactors)
  (f) process pumping maintenance materials cost (pressure contactors)
  (g) labor cost for onsite GAC transport
  (h) water cost for onsite GAC transport
  (i) overhead cost on all labor

• reactivation capital cost
  (a) construction cost of reactivation system
  (b) capital overhead cost plus engineering fee

• reactivation O&M cost
  (a) process electrical cost
  (b) building electrical cost
  (c) maintenance materials cost
  (d) O&M labor cost
  (e) process fuel cost
  (f) GAC makeup cost
  (g) process water cost
  (h) overhead cost on all labor

• offsite GAC transport capital cost
  (a) cost of transport vehicles

• offsite GAC transport O&M cost
  (a) transport labor cost (loading, haul, and off-loading)
  (b) transport fuel cost
  (c) maintenance materials cost for vehicles

Table 13. Capital Cost Function Coefficients and Exponents

Process Name	USRT Range	a	b	c
Steel pressure contactors	390–2260 ft^{3a}	80,475	106.67	0.919
Concrete gravity contactors	350–10,600 ft^{3a}	0	16418.64	0.359
In-plant pumping	1–10 mgd^b	25,429	2531.07	1.334
	10–200 mgd^b	50,163	2390.25	1.134
Onsite GAC storage	1,000–20,000 ft^{3a}	14,106	6.69	1.098
Initial cost of GAC	10,000–10,000,000 lb^c	0	—	0.933
Fluid bed reactivation	6000–50,000 lb/day^d	599,030	4694.80	0.494
Infrared reactivation	2400–50,000 lb/day^d	222,261	49.52	0.933
Multihearth reactivation	27–47 ft^{2e}	521,081	3.13 × 10^{-6}	6.726
	47–1510 ft^{2e}	898,670	6806.69	0.771

^{a}To convert from ft^3 to m^3 multiply by 0.02832.
^{b}To convert from mgd to m^3/day, multiply by 3785.
^{c}To convert from lb to kg, multiply by 0.37324.
^{d}To convert from lb/day to kg/day, multiply by 0.37324.
^{e}To convert from ft^3 to m^3 multiply by 0.092903.

Capital Cost Equations

The form of the capital cost equations was as follows:

$$CC = [a + b(USRT)^c](NU)(CCI/2850.7) + (OHEF) \qquad (1)$$

$$AC = (CC) [I(1 + I)^N/((1 + I)^N - 1)] \qquad (2)$$

where CC = construction cost, dollars
    AC = amortized capital cost, \$/year
    a,b,c = equation coefficients and exponent
    USRT = design variable
    NU = number of units in process (i.e., NU = 1 for a single contactor or reactivator)
    CCI = ENR construction cost index, base year 1913
    OHEF = overhead, interest during construction, engineering fee, dollars (must be added separately)
    I = period interest rate (decimal)
    N = payback period, years

Table 13 contains the coefficients for the capital cost equations.

The assumptions used in developing the capital cost equations are as follows:

1. Contactor capital cost is based on the effective GAC volume (ft^3) of an individual contactor.
2. Pump costs are based on the capacity of an individual pump (mgd); the capital cost includes a standby pump.
3. The onsite GAC storage cost is based on total storage volume (ft^3); one unit is assumed.
4. Given the cost of GAC (dollars) for a specified amount (pounds), the

coefficient "b" in the equation "cost = b(pounds)$^{0.933}$" can then be used to estimate cost of GAC for different amounts. For example, in this report \$0.90/lb GAC cost at 10,000,000 lb was assumed; therefore "b" is 2.65; GAC cost (\$) = 2.65(pounds GAC)$^{0.933}$. The value for NU is assumed equal to 1. The general equation is as follows: GAC cost (\$) = (\$/lb)(xlb)$^{0.04}$(lb)$^{0.96}$ where unit purchase price (\$/lb) is based on a specified quantity of carbon (xlb) in pounds and (lb) is the amount of carbon in pounds for which the cost estimate is being determined.

5. Costs for fluid bed and infrared furnaces are based on the pounds per day rated capacity for a single reactivator.
6. Costs for multihearth furnace are based on effective hearth area (ft^2) for a single reactivator.

O&M Cost Equations

The O&M cost equation forms are as follows:

$$PE = [d + e(USRT)^f](NU)(\$/kWh) \tag{3}$$

$$PEI = [d + e(USRT)^f](TDH/35)(NU)(\$/kWh) \tag{4}$$

$$BE = [g + h(USRT)^i](NU)(\$/kWh) \tag{5}$$

$$MAINT = [j + k(USRT)^\ell](PPI/199.7)(NU) \tag{6}$$

$$MAINTO = [j + k(USRT)^\ell(MILES)^{\ell\ell}](PPI/199.7) \tag{7}$$

$$LABOR = [m + n(USRT)^o](NU)(\$/hour) \tag{8}$$

$$LABORO = [m + n(USRT)^o(MILES)^{oo}](\$/hour) \tag{9}$$

$$FUELG = [p + q(USRT)^r](NU)(\$/ft^3) \tag{10}$$

$$FUELO = [p + q(USRT)^r](1/139)(NU)(\$/gal) \tag{11}$$

$$FUELD = [p + q(USRT)^r(MILES)^{rr}](\$/gal) \tag{12}$$

$$MAKEUP = [(XLBS)^t(USRT)^u](\$/lb) \tag{13}$$

where PE = process electrical cost, \$/year
   PEI = electrical cost for in-plant pumping, \$/year
   BE = building electrical cost, \$/year
   MAINT = maintenance materials cost for onsite systems, \$/year
   MAINTO = maintenance materials cost for offsite transportation, \$/year
   LABOR = process O&M labor cost, \$/year
   LABORO = labor cost for offsite GAC transport, \$/year
   FUELG = process natural gas cost, \$/year
   FUELO = fuel oil cost, \$/year
   FUELD = transport diesel cost, \$/year
   MAKEUP = carbon makeup cost, \$/year when USRT is in lb/year
   d,e,f,...t,u = equation coefficients and exponents
   USRT = design variable

MILES = one-way distance, miles
TDH = total dynamic head, ft
XLBS = pounds of carbon on which the purchase cost ($/lb) is based
NU = number of units in process. This variable is applicable only to in-plant pumping, multihearth reactivation, infrared reactivation, and fluid bed reactivation. NU = 1 for the other processes, i.e., contactors and offsite transport.
PPI = Producers' Price Index value

Table 14 contains the O&M cost function coefficients and exponents. The following assumptions were made regarding the O&M cost equation derivation:

1. All O&M costs are in dollars per year and assume utilization 100% of the time. If a process is used part of the time, then the cost estimate should be adjusted according to percent utilization.
2. O&M cost for contactors is based on total filter cross-sectional area of all contactors combined.
3. Pump cost is based on capacity of an individual pump in millions of gallons per day.
4. Offsite GAC transport cost is based on pounds of *dry*-weight GAC hauled. If GAC is wet-drained and road weight limits exist, then the number of truckloads may double, thus doubling the estimates.

The following O&M cost assumptions are based on the CWW and MWW GAC treatment experience:

1. Labor for onsite GAC transport is 0.383 man-hours/1000 lb of GAC transferred.
2. Water for onsite GAC transport is 2 gal/lb of GAC transferred.
3. Water for a reactivation process is assumed to be 21 gal/lb of GAC reactivated.

ESTIMATING *GAC* SYSTEM COSTS

The information obtained in the CWW and MWW research projects provided many details and insights concerning the cost of operating GAC treatment systems. However, the CWW and MWW data are site-specific and alone cannot provide a basis for evaluating GAC systems of different sizes and types. Therefore, cost estimates for various GAC treatment scenarios have been generated to examine the effect of economies of scale in GAC systems and to evaluate the cost tradeoffs between alternative GAC treatment systems.[6]

The cost estimates presented in this paper were generated using the cost estimating functions that were discussed earlier. Each of the cost functions represents an individual cost item associated with a treatment process, and incorporates one or more design variables and cost parameters. For exam-

Table 14. O&M Cost Function Coefficients and Exponents

Process Name	USRT Range	Model Value
Steel pressure contactors	157 – 6,790 ft²[a]	d = 0; e = 5.92; f = 0.999; g = 146831; h = 130.19; i = 1.093; j = 774; k = 5.07; ℓ = 1.007; m = 1572; n = 12.57; o = 0.698
Concrete gravity contactors	140 – 28,000 ft²[b]	d = 0; e = 4.83; f = 1.000; g = 10831; h = 361.31; i = 0.912; j = 109; k = 17.99; ℓ = 0.744; m = 701; n = 3.25; o = 0.838
In-plant pumping	1 – 10 mgd[c]	d = 0; e = 52467.0; f = 0.999; j = 263; k = 86.71; ℓ = 1.188; m = 476; n = 44.13; o = 0.777
In-plant pumping	10 – 200 mgd[c]	d = 0; e = 52465.0; f = 1.000; j = 0; k = 341.54; ℓ = 0.812; m = 574; n = 12.60; o = 1.112
Offsite GAC transport (< 25 miles)	30,000 – 3,000,000 lb[b]	j = 0; k = 2.31×10^{-5}; ℓ = 0.992; $\ell\ell$ = 0.998; m = 0; n = 3.15×10^{-5}; o = 1.067; oo = 0.477; p = 0; q = 2.38×10^{-5}; r = 0.989; rr = 0.984
Offsite GAC transport (> 25 miles)	30,000 – 3,000,000 lb[b]	j = 0; k = 2.16×10^{-5}; ℓ = 1.001; $\ell\ell$ = 0.998; m = 0; n = 1.91×10^{-4}; o = 1.021; oo = 0.131; p = 0; q = 1.83×10^{-5}; r = 1.000; rr = 1.012
Fluid bed reactivation	6,000 – 60,000 lb/day[d]	d = 0; e = 43.80; f = 1.000; j = 7232; k = 385.19; ℓ = 0.353; m = 4000; n = 245.00; o = 0.450; p = 111,107; q = 1046.14; r = 1.008
Infrared reactivation	2,400 – 60,000 lb/day[d]	d = 44.779; e = 315.55; f = 0.988; g = 417; h = 23.14; i = 0.753; j = 0; k = 427.25; ℓ = 0.397; m = 4000; n = 25.7; o = 0.742
Multihearth reactivation	27 – 1,510 ft²[a]	d = 472,706; e = 8517.36; f = 0.755; g = 11,246; h = 312.08; i = 0.649; j = 0; k = 846.15; ℓ = 0.401; m = 3000; n = 604.0; o = 0.728; p = 366,525; q = 287,714.93; r = 0.899
Makeup carbon	10,000 – 10,000,000 lb[b]	t = 0.04; u = 0.96

[a]To convert from ft² to m², multiply by 0.092903.
[b]To convert from lb to kg, multiply by 0.37324.
[c]To convert from mgd to m³/day, multiply by 3785.
[d]To convert from lb/day to kg/day, multiply by 0.37324.

ple, the total O&M cost of a fluid bed reactivation system is estimated by generating and summing the values from seven individual fluid bed cost functions (i.e., reactivation labor, process and building electric, maintenance materials, process fuel, process water, makeup GAC, and labor overhead).

The costs associated with two types of GAC adsorption systems were examined: deep bed steel pressure GAC contactors and deep bed concrete gravity GAC contactors. The cost associated with three types of onsite reactivation units were considered: fluid bed, infrared, and multihearth furnaces. An offsite reactivation scenario was also evaluated.

The cost of a concrete gravity contactor system includes the following: contactor system construction; virgin GAC storage construction; construction overhead; initial GAC; contactor system electric, maintenance material, and labor; labor for onsite carbon transport; overhead on labor; and water for carbon transport. The cost of a pressure contactor system includes the same items with the addition of the capital, electric, labor, and maintenance material for process pumping. The cost of an onsite reactivation alternative includes the following: reactivation system capital plus overhead; process and building electric; maintenance material; labor; fuel; GAC makeup; water for process; and overhead on labor. An offsite carbon transport alternative includes the capital cost for the vehicle(s), the cost of labor for loading and off-loading carbon, fuel cost, and maintenance material cost for the truck(s), assuming the transport of wet-drained carbon. Cost parameters and assumptions used in this analysis are presented in Table 15.

GAC Treatment Scenarios

Cost estimates for several treatment systems of different sizes and types have been generated to examine the differences in cost that exist as a result of variations in system characteristics. Systems with flow capacities of 3785, 11,355, 18,925, 37,850, 94,625, 189,250, 378,500, and 662,375 m³/day (1, 3, 5, 10, 25, 50, 100, and 175 mgd) were evaluated, each having an EBCT of 15 min, bed depths of about 3.35 to 3.66 m (11 to 12 ft), and GAC reactivation every three months. The cost estimates include capital and O&M expenses associated with steel pressure and concrete gravity contactors and onsite fluid bed, infrared, and multihearth GAC reactivation. A unit cost summary comparing all the alternatives is presented in Table 16. Tables 17–23 contain design assumptions for various scenarios.

Adsorption Alternatives

The effect of economies of scale associated with the GAC contactor alternatives (adsorption excluding reactivation) is shown in Figure 3. The smaller systems, $\leq$ 37,850 m³/day ($\leq$ 10 mgd), experience significantly higher unit costs than the larger systems. For the steel pressure contactors, the unit cost varies about 6.1 ¢/m³ (23 ¢/1000 gal) from the 3785-m³/day

Table 15.   Parameters Used in Computing Cost Estimates

Parameter	Value
GAC cost	3,732,400 kg @ $2.64/kg
	(10,000,000 lb @ $0.90/lb)
O&M labor cost	$9/hr
Electric power	$0.07/kWh
Fuel oil	$317.0/m³
	($1.20 gal)
Natural gas	$0.1236/m³
	($0.0035/ft³)
Process and transport water	$0.13/m³
	($0.50/1000 gal)
Capital recovery interest rate	10%
Amortization period	15 years
Labor overhead	75%
Construction cost index for 1984	4131
Producers' Price Index for 1984	302
GAC density	480.6 kg/m³
	(30 lb/ft³)
Volume of GAC per mgd capacity of system	900 m³/m²
	(1392 ft³/mgd)
Reactivation uptime factor	70%
Multihearth loading rate	244.12 kg/m²/day
	(50 lb/ft²/day)
GAC loss, reactivation, and onsite transport	15%
GAC loss, offsite transport plus reactivation	18%
Capital overhead with engineering fee	13.5%

(1-mgd) system to the 94,625-m³/day (25-mgd) system while there is only a 0.26-¢/m³ (1 ¢/1000 gal) unit cost variation from 94,625 m³/day (25 mgd) to 662,375 m³/day (175 mgd). For the concrete gravity contactors, the unit cost varies about 4 ¢/m³ (15 ¢/1000 gal) from 3785 m³/day (1 mgd) to 94,625 m³/day (25 mgd), while the variation in cost is about 0.26 ¢/m³ (1 ¢/1000 gal) from 94,625 m³/day (25 mgd) to 662,375 m³/day (175 mgd). Comparing the alternative adsorption systems, it appears that the steel pressure contactor systems exhibit higher costs (about 6 to 14 ¢/1000 gal; 1.5 to 3.7 ¢/m³) than the concrete gravity contactor systems in all cases. The concrete gravity systems exhibit lower capital and O&M costs than the steel pressure systems as a result of the greater economies of scale inherent in a smaller number of larger-sized concrete contactors compared to a larger number of smaller-sized steel contactors.

Table 16. Unit Cost Estimates for Various GAC Systems

Item	GAC System (mgd)[a]							
	1	3	5	10	25	50	100	175
Pressure contactors (¢/1000 gal)[b]	35.20	19.83	17.75	16.15	11.94	11.32	10.97	10.87
Concrete gravity contactors (¢/1000 gal)[b]	21.16	11.27	8.76	8.16	5.70	4.98	4.77	4.58
Fluid bed reactivation (¢/lb)[c]	—	85.64	64.22	46.34	33.50	28.00	27.30	25.99
Infrared reactivation (¢/lb)[c]	—	64.27	53.55	44.20	36.78	33.09	32.39	31.33
Multihearth reactivation (¢/lb)[c]	—	91.92	81.35	60.84	45.85	39.04	38.35	36.76
Virgin replacement (¢/lb)[c]	118.38	109.98	106.28	101.46	95.42	91.08	86.96	83.75
Pressure contactors								
with fluid bed	—	59.04	47.15	37.36	27.26	24.14	23.46	22.76
with infrared	—	49.25	42.27	36.38	28.76	26.48	25.79	25.20
with multihearth	—	61.91	54.99	44.00	32.91	29.20	28.51	27.69
Concrete gravity contactors								
with fluid bed	—	50.48	38.16	29.37	21.02	17.80	17.26	16.47
with infrared	—	40.69	33.28	28.39	22.52	20.14	19.59	18.92
with multihearth	—	53.35	46.00	36.01	26.67	22.86	22.31	21.40

[a] mgd × 3,785 = m³/day.
[b] ¢/1000 gal × 0.2642 = ¢/m³.
[c] ¢/lb × 2.2 = ¢/kg.

COST ANALYSIS FOR *GAC*   325

Table 17. Design Parameters for Steel Pressure Contactor Systems Scenario

	Plant Capacity (mgd)[a]							
	1	3	5	10	25	50	100	175
Number of contactors	2	3	5	10	10	19	37	65
Hydraulic loading (gpm/ft²)[b]	5.46	6.14	6.14	6.14	5.52	5.82	5.97	5.95
Diameter (ft)[c]	9	12	12	12	20	20	20	20
Area of each (ft²)[d]	63.62	113.10	113.10	113.10	314.16	314.16	314.16	314.16
GAC depth (ft)	10.95	12.31	12.31	12.31	11.07	11.67	11.97	11.93
Volume of GAC (ft³)[e] per contactor	696.64	1,392.26	1,392.26	1,392.26	3,477.75	3,666.25	3,760.50	3,747.93
EBCT (min)	15	15	15	15	15	15	15	15
Total carbon loss (%)	15	15	15	15	15	15	15	15
Reactivation frequency (no./year)	4	4	4	4	4	4	4	4
Reactivation demand (lb/year)[f]	167,194	501,214	835,356	1,670,712	4,173,300	8,359,050	16,696,620	29,233,854
Size of each furnace (lb/day)[g]	—	1,962	3,270	6,539	16,335	32,716	32,674	38,139
Number of furnaces	0	1	1	1	1	1	2	3
Virgin storage size (lb)[h]	—	2,000	3,500	6,500	16,000	33,000	66,000	120,000

[a] mgd × 3785 = m³/day.
[b] gpm/ft² × 40.7 = L/min/m².
[c] ft × 0.30480 = m.
[d] ft² × 0.092903 = m².
[e] ft³ × 0.02932 = m³.
[f] lb/year × 0.37324 = kg.
[g] lb/day × 0.37324 = kg/day.
[h] lb × 0.37324 = kg.

Table 18.   Design Parameters for Concrete Gravity Systems Scenario

	Plant Capacity (mgd)[a]								
	1	3	5	10	25	50	100	175
Number of contactors	1	1	1	2	2	3	6	10
Hydraulic loading (gpm/ft²)[b]	5.46	6.14	6.14	6.14	5.52	5.82	5.97	5.95
Area per contactor (ft²)[c]	127.2	339.3	565.5	565.5	1,570.8	1,989.7	1,937.3	2,042.0
GAC depth (ft)[d]	10.95	12.31	12.31	12.31	11.07	11.67	11.97	11.93
Volume of GAC per contactor (ft³)[e]	1,392.8	4,176.8	6,961.3	6,961.3	17,388.76	23,219.8	23,189.5	24,361.1
EBCT (min)	15	15	15	15	15	15	15	15

Note: Reactivation requirements same as pressure contactors.

[a]mgd $\times$ 3785 = m³/day.
[b]gpm/ft² $\times$ 40.7 = L/min/m².
[c]ft² $\times$ 0.092903 = m².
[d]ft $\times$ 0.30480 = m.
[e]ft³ $\times$ 0.02832 = m³.

Table 19. Steel Pressure Contactor Costs

| | \multicolumn{7}{c|}{System (mgd)[a] Cost ($/year)} |
	1	3	5	10	25	50	100	175
Process and building electric	15,823	26,740	38,136	67,206	163,926	318,464	635,299	1,129,699
Maintenance material	2,709	4,762	6,988	12,699	33,705	62,185	118,060	203,299
O&M labor	22,160	25,970	29,255	36,116	54,611	76,984	116,147	170,048
Labor overhead	17,073	20,774	24,101	31,406	51,749	79,349	130,275	203,113
Labor for GAC transfer	576	1,728	2,879	5,759	14,385	28,814	57,553	100,769
Transfer water	167	501	835	1,671	4,173	8,359	16,697	29,234
Amortized capital	69,996	136,753	221,869	434,716	767,269	1,492,456	2,931,487	5,107,218
TOTAL COST	128,484	217,228	324,064	589,573	1,089,816	2,066,610	4,005,517	6,943,309

[a]mgd × 3785 = m³/day.

Table 20. Concrete Gravity Contactor Costs

	System (mgd)[a] Cost ($/year)							
	1	3	5	10	25	50	100	175
Process and building electric	2,903	6, 012	9,138	16,549	40,939	73,019	133,687	223,313
Maintenance material	1,166	2,242	3,202	5,252	11,043	17,700	28,958	43,952
O&M labor	8,006	10,170	12,234	16,900	31,240	49,000	80,936	125,970
Labor overhead	6,437	8,924	11,335	16,994	34,219	58,360	103,867	170,054
Labor for GAC transfer	576	1,728	2,879	5,759	14,385	28,814	57,553	100,769
Transfer water	167	501	835	1,671	4,173	8,359	16,697	29,234
Amortized capital	57,979	93,900	120,418	234,872	384,470	674,151	1,319,764	2,233,059
TOTAL COST	77,235	123,477	160,042	297,996	520,469	909,403	1,741,460	2,926,350

[a] mgd × 3785 = m³/day.

Table 21. Onsite Fluid Bed Reactivation Costs

	System (mgd)[a] Cost ($/year)						
	3	5	10	25	50	100	175
Process and building electric	4,210	7,017	14,034	35,056	70,216	140,252	245,564
Maintenance material	13,579	14,749	16,719	20,176	23,656	47,297	73,637
O&M labor	71,986	84,079	105,643	146,649	191,218	382,244	609,241
Labor overhead	53,990	63,059	79,232	109,987	143,414	286,683	456,931
Process fuel	13,849	22,724	45,021	112,267	225,441	450,301	789,601
Process water	5,263	8,771	17,542	43,820	87,770	175,315	306,955
Makeup GAC	93,899	151,233	288,740	678,339	1,296,915	2,473,156	4,170,717
Amortized capital	172,503	184,837	207,414	251,947	302,058	603,897	946,913
TOTAL COST	429,279	536,469	774,344	1,398,240	2,340,688	4,559,144	7,599,020

[a]mgd $\times$ 3785 = m^3/day.

Table 22. Onsite Infrared Reactivation Costs

	System (mgd)[a] Cost ($/year)						
	3	5	10	25	50	100	175
Process and building electric	30,406	48,816	94,424	229,437	452,830	904,519	1,579,164
Maintenance material	9,174	11,236	14,797	21,282	28,039	56,051	89,400
O&M labor	70,111	90,811	134,944	241,663	387,637	774,582	1,293,961
Labor overhead	52,583	68,108	101,208	181,247	290,727	508,936	970,470
Process water	5,263	8,771	17,542	43,820	87,770	175,315	306,955
Makeup GAC	93,899	151,233	288,740	678,339	1,296,915	2,473,156	4,170,717
Amortized capital	60,704	68,423	86,930	139,375	222,640	444,863	748,494
TOTAL COST	322,139	447,398	738,585	1,535,163	2,766,559	5,409,421	9,159,162

[a]mgd × 3785 = m³/day.

Table 23. Onsite Multihearth Reactivation Costs

	System (mgd)[a] Cost ($/year)						
	3	5	10	25	50	100	175
Process and building electric	30,846	34,075	41,004	57,896	81,192	162,273	264,578
Maintenance material	3,901	4,788	6,322	9,127	12,059	24,106	38,472
O&M labor	73,920	98,706	151,096	276,327	445,753	890,708	1,488,526
Labor overhead	55,440	74,029	113,322	207,245	334,315	668,031	1,116,394
Process fuel	49,304	76,750	141,209	318,748	593,270	1,185,175	2,041,951
Process water	5,263	8,881	17,542	43,820	87,770	175,315	306,955
Makeup GAC	93,899	151,233	288,740	678,339	1,296,915	2,473,156	4,170,717
Amortized Capital	148,153	231,286	257,386	322,051	412,527	824,621	1,319,748
TOTAL COST	460,726	679,638	1,016,623	1,913,553	3,263,801	6,403,384	10,747,342

[a]mgd × 3785 = m³/day.

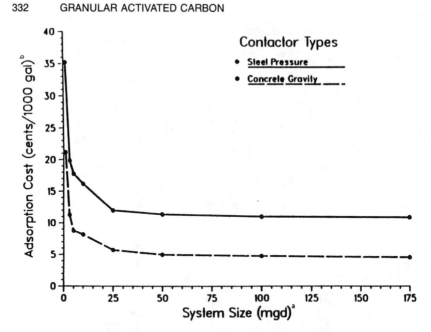

Figure 3. GAC contactor cost. (a. mgd × 3785 = m³/day; b. ¢/1000 gal × 0.264 = ¢/m³.)

Onsite Reactivation and Replacement Alternatives

The onsite reactivation alternatives exhibit economies of scale similar in nature to those of GAC contactors with systems smaller than 94,625 m³/day (25 mgd) experiencing significantly higher unit costs than the larger systems (Figure 4). Infrared reactivation appears to be the least expensive alternative for smaller plants (< 15 mgd; 56,775 m³/day in this scenario), while fluid bed reactivation is cheaper for larger systems. Regardless of which technology is used, Table 13 shows that for this scenario, onsite reactivation is cheaper than replacement of spent carbon with virgin GAC. Even with a 11,355-m³/day (3-mgd) system, onsite reactivation appears to be more cost-effective.

The cost tradeoffs between onsite reactivation and replacement of spent carbon with virgin GAC are shown in Figure 5. Figure 5 presents the total costs (dollars/year) of single fluid bed, infrared, and multihearth furnace systems compared to complete GAC replacement, as a function of the quantity of carbon to be reactivated or replaced. The reactivation costs include increased furnace capacity to incorporate a 70% uptime operating factor. The figure shows that even for very small amounts of carbon, i.e., 136,200 kg/year (300,000 lb/year), a small infrared reactivator (rated 50 lb/hr; 22.7 kg/hr) appears to be more economical than replacement of spent carbon with virgin GAC. For larger amounts of carbon, onsite reactivation becomes increasingly more cost-effective.

The cost tradeoffs among the three types of reactivation furnaces are

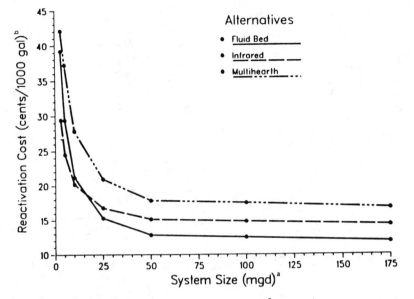

Figure 4. GAC reactivation cost. (a. mgd × 3785 = m³/day; b. ¢/1000 gal × 0.264 = ¢/m³.)

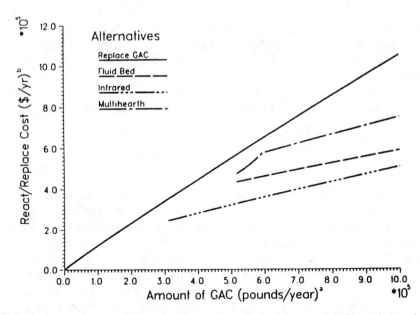

Figure 5. GAC reactivation and replacement costs. (a. lb/year × 0.37324 = kg/year.)

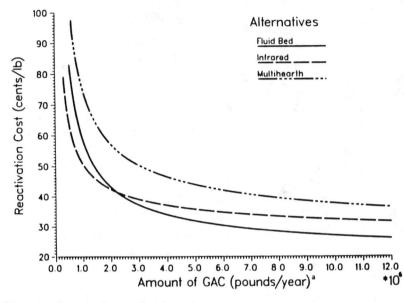

Figure 6.   Costs for three types of reactivators. (a. lb/year × 0.37324 = kg/year.)

shown in Figure 6. This figure presents the unit costs of single individual fluid bed, infrared, and multihearth reactivators as a function of the amount of carbon to be reactivated. Infrared reactivation appears to be the most economical reactivation alternative for amounts of carbon less than about 908,000 kg/year (2,000,000 lb/year). For quantities of GAC greater than 3178–4086 kg/day (7000–9000 lb/day) fluid bed reactivation is the most economical choice. This breakeven point between infrared and fluid bed reactivation, however, is highly dependent on factors such as labor rates, fuel rates, electrical power rates, assumed uptime operating factors, etc.

Total Cost of Various GAC Treatment Systems

The costs presented in Table 16 assume an EBCT of 15 min and a reactivation frequency of every three months. These costs may vary significantly with variations in EBCT, reactivation frequency, hydraulic loading, size of contactors, and other system parameters. As an example, Figures 7 and 8 present the total GAC treatment unit cost (adsorption plus onsite reactivation) associated with concrete gravity and steel pressure contactor systems for 37,850-m³/day (10 mgd) and 378,500-m³/day (100 mgd) plants, as a function of reactivation frequency, with EBCTs of 15 and 20 min. In general, the figures show the following:

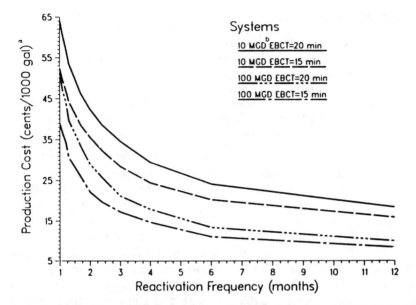

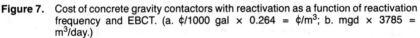

Figure 7.   Cost of concrete gravity contactors with reactivation as a function of reactivation frequency and EBCT. (a. ¢/1000 gal × 0.264 = ¢/m³; b. mgd × 3785 = m³/day.)

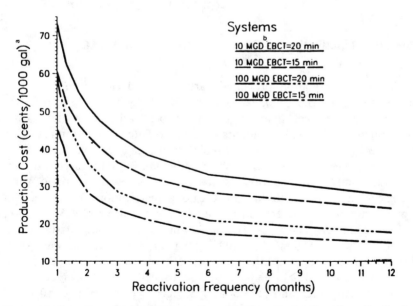

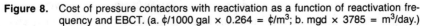

Figure 8.   Cost of pressure contactors with reactivation as a function of reactivation frequency and EBCT. (a. ¢/1000 gal × 0.264 = ¢/m³; b. mgd × 3785 = m³/day.)

1. The 378,500-m³/d (100 mgd) plant exhibits lower unit water production costs than the 37,850-m³/d (10 mgd) plant because of the greater economies of scale inherent in larger facilities.
2. Systems requiring more frequent reactivation for a given plant size exhibit higher unit water production costs.
3. The unit water production costs associated with the systems having the higher EBCT values (i.e., 20 min) for a given plant at a fixed reactivation frequency are higher, because larger facilities are required and larger volumes of carbon must be reactivated, thus increasing the costs.

The types of cost curves shown in Figures 7 and 8 can be generated using one's site-specific system parameters and assumptions to examine the cost tradeoffs among alternative GAC systems of different EBCTs. For example, one may want to determine for a 378,500-m³/day (100-mgd) concrete gravity contactor system whether an EBCT of 15 min is more economical than an EBCT of 20 min, given a fixed water quality treatment goal. If it is determined that the operating service life of the GAC in a contactor with an EBCT of 15 min is 90 days, Figure 7 shows that the total system unit cost is about 4.5 ¢/m³ (17 ¢/1000 gal). In order for the system with an EBCT of 20 min to be more cost-effective than one with a 15-min EBCT, the operating service life must be approximately 130 days or longer (4.25 months, from Figure 7). If it is determined that the system with 20 min EBCT provides a service life of 160 days, then this system is more economical and cost-effective than the system with the 15-min EBCT. These types of cost curves are useful in estimating the cost-breakeven points for different alternatives.

Regional Reactivation vs Onsite Reactivation and GAC Replacement

The cost information presented earlier showed that even for small amounts of GAC (i.e., 300,000 lb/year; 136,200 kg/year) onsite reactivation is more economical than replacement of spent carbon with virgin GAC. For systems using smaller amounts of GAC, offsite reactivation may be a cost-effective alternative to virgin replacement, and regional reactivation may be more economical than onsite reactivation for some larger utilities.

As an example, the scenario presented earlier for several utilities of varying size is extended here to evaluate a regional reactivation alternative. This regional reactivation system consists of the 3785-, 11,855-, 18,925-, 37,850-, and 94,625-m³/day (1-, 3-, 5-, 10-, and 25-mgd) plants using offsite reactivation at the 378,500-m³/day (100 mgd) plant site. The one-way travel distance for each of the utilities using offsite transport is 121 km (75 miles). Four trucks are utilized, each having a haul capacity of 9080 kg (20,000 lbs) for wet-weight GAC, and each costing $75,000 depreciated over 10 years. Reactivation demands are based on a reactivation frequency of every three months. Total GAC losses for the utilities using offsite transport are 18%,

Table 24. Costs Associated with Regional Reactivation Scenario

System (mgd)[a]	GAC Amount (lb/year)[b]	Transport O&M ($/year)	Transport Capital ($/year)	Makeup GAC ($/year)	Furnace O&M and Capital ($/year)	Total Cost ($/year)	Unit Cost (¢/lb)[c]
1	167,194	3,679	683	39,965	18,062	62,389	37.31
3	501,211	11,190	2,046	111,311	54,148	178,695	35.65
5	835,356	18,777	3,411	179,275	90,246	291,709	34.92
10	1,670,712	37,906	6,821	342,281	180,492	567,500	33.96
25	4,173,300	95,863	17,039	804,123	450,854	1,367,879	32.77
100	6,969,620	0	0	2,473,156	1,803,786	4,276,942	25.61
Total GAC	24,044,396						

[a] mgd × 3785 = m³/day.
[b] lb/year × 0.454 = kg/year.
[c] ¢/lb × 2.2 = ¢/kg.

Table 25. Unit Cost Summary of Regional Alternatives in Scenario

	GAC System (mgd)[a]					
	1	3	5	10	25	100
	Cost (¢/lb of GAC)[b]					
Regional reactivation	37.31	35.65	34.92	33.96	32.77	25.61
Onsite reactivation	—	64.27	53.55	44.20	33.50	27.30
Replacement, virgin GAC	118.38	109.98	106.28	101.46	95.42	86.96

[a] mgd × 3785 = m³/day.
[b] ¢/lb × 2.2 = ¢/kg.

and the 37,850-m³/day (100 mgd) plant (onsite) assumes 15% GAC losses. Two fluid bed reactivators, each having a capacity of 21,565 kg/day (47,500 lb/day), are utilized at the 378,500-m³/day (100 mgd) plant. The annual costs associated with the regional reactivation scenario are presented in Table 24. Table 25 presents a unit cost summary comparing the costs for regional reactivation, onsite reactivation, and replacement with virgin GAC for each of the utilities.

REFERENCES

1. Miller, R., and D. J. Hartman. "Feasibility Study of Granulated Activated Carbon Adsorption and On-Site Regeneration," Volume 1, U.S. EPA Report 600/2-82-087A (1982).
2. Gumerman, R., R. Culp, and S. Hansen. "Estimating Water Treatment Costs, Vol. II, Cost Curves Applicable to 1 to 200 mgd Treatment Plants," U.S. EPA Report 600/2-79-162b (1979).
3. Koffskey, W. E., and B. W. Lykins, Jr. "Granular Activated Carbon Adsorption with On-Site Infrared Furnace Reactivation," U.S. EPA Report 600/S2-88/058 (1989).
4. Adams, J. Q., R. M. Clark, B. W. Lykins, Jr., and D. Kittridge. "Regional Reactivation of Granular Activated Carbon," *J. Am. Water Works Assoc.* 78(5):38-41 (1986).
5. Kittredge, D., R. Beaurivage, and D. Paris. "Granular Activated Carbon Adsorption and Fluid-Bed Reactivation at Manchester, New Hampshire," U.S. EPA Report 600/2-83-104, (1984).
6. Gumerman, R. C., R. L. Culp, and S. P. Hansen. "Estimating Water Treatment Cost, Volume I—Summary" (Cincinnati, OH: U.S. EPA).
7. Clark, R. M., and P. Dorsey. "A Model of Costs for Treating Drinking Water," *J. Am. Water Works Assoc.* 74 (12):618-627 (1982).
8. Kittredge, D., R. Beaurivage, D. Paris, B. Lykins, Jr., and J. DeMarco. "Granular Activated Carbon Adsorption and Fluid-Bed Reactivation at Manchester, New Hampshire," U.S. EPA Report 600/2-83-104 (1982).
9. Lykins, Jr., B., E. Geldreich, J. Adams, J. Ireland, and R. Clark. "Field Scale Granular Activated Carbon Studies for Removal of Organic Contaminants Other Than Trihalomethanes from Drinking Water," U.S. EPA Report 600/2-84-165 (1984).
10. McGuire, M., and I. Suffet. "Treatment of Water by Granular Activated Carbon," American Chemical Society (1983).
11. Culp, R., J. Faisst, and C. Smith. "Granular Activated Carbon Installations," U.S. EPA Report 600/2-81-177 (1981).

Index